D1006172

The Number of the Heavens

THE
NUMBER
OF THE
HEAVENS

A History of the Multiverse and the Quest to Understand the Cosmos

TOM SIEGFRIED

Harvard University Press

Cambridge, Massachusetts
London, England
2019

Library of Congress Cataloging-in-Publication Data
is available from loc.gov

ISBN: 978-0-674-97588-0 (alk. paper)

Contents

Preface

After his conquest of most of the known world, Alexander the Great wept, legend has it, because there were no more worlds to conquer. But actually, as the black-hatted villain played by Ed Harris informed us in season two of *Westworld*, Alexander cried when told that there was an infinity of worlds—and he hadn't yet finished conquering one of them.

The popular, erroneous story of Alexander's sorrow is a misquotation of the Greek-Roman scholar and biographer Plutarch. He reported that the existence of a multiplicity of worlds was revealed to Alexander by the philosopher Anaxarchus, a follower of Democritus, the philosopher of the fifth century BC who taught that everything is made of atoms. Democritus also believed that the known cosmos—believed in Alexander's time to be a set of spheres carrying planets and stars around the Earth—was just one of countless many. Anaxarchus contradicted Alexander's earlier teacher, Aristotle, who argued with Vulcan-like assuredness that logic prohibited the existence of multiple worlds.

In the view of most ancient and early medieval philosophers, Aristotle won the argument. But following an edict contradicting Aristotle, issued by the bishop of Paris in 1277, medieval philosophers revisited the possibility of a plurality of worlds. Scholars continued to debate that proposition over the centuries that followed. The debate never died, but its terms continually shifted as scientists revised their notion of what constitutes the "world," or the universe. From the sun-centered solar system described by Copernicus, to the vast "universe" of stars known as the Milky

Way galaxy, to the expansive and expanding Einsteinian spacetime filled by billions and billions of such galaxies, the definition of *universe* kept changing, requiring revision every time scientists discovered there was more than one. In the late twentieth century, the debate reignited as new data and theories about the cosmos led many experts to suspect that our universe is not alone. Today cosmological authorities argue once again about whether there is one universe or many—a multiverse.

This book tells the history of this eternal debate, from the time of the ancient Greeks to the present. My interest in this issue originated during my readings of medieval scientific history, when I realized that the arguments against the multiverse today echo in many respects the arguments by medieval natural philosophers opposed to the possibility of a plurality of worlds. Those parallels, I soon learned, extended back also to the ancient Greeks. The multiverse debate is like a long-running TV series with new characters replacing the old ones but reprising common themes.

What follows is my account of the history of the multiverse debate, from antiquity to the present. I make no attempt, though, to survey the wide range of ongoing research today relevant to the multiverse issue. That would take another whole book.

The first part of my story naturally relies on the writings of the most important philosophers and scientists who have engaged with this issue as well as the analyses by modern scholars who have studied and assessed the debates of the past. Most helpful in this regard have been the writings of Pierre Duhem, Alexandre Koyré, David Lindberg, Michael Crowe, A. C. Crombie, Steven Dick, and especially Edward Grant. In the later chapters, I've relied on my own reporting of cosmological developments over the last 35 years, as science editor of the *Dallas Morning News*, a contributor to *Science* magazine, and as managing editor and editor in chief of *Science News* magazine.

During these years I have benefited enormously from the co-operation of a great many scientists—too many to name here, but most will appear in the pages that follow. I must mention one of the most helpful, though, who sadly will not see this book: Joe Polchinski, of the University of California, Santa Barbara, who died in 2018.

Also too numerous to mention are my many friends, colleagues, and relatives who have listened patiently to my undigested thoughts on this topic in recent years and posed polite questions and comments, often very insightful. I owe special thanks to my former *Science News* colleague Beth Rakouskas for creating the figures in chapters 2 and 13.

Thanks are also due to my editor at Harvard University Press, Jeff Dean, for his many apt observations about the manuscript and to several reviewers for their very helpful comments. Extra appreciation is extended to Steve Maran, former press officer for the American Astronomical Society, who first approached me about the prospect of writing what turned out to be this book. And it would be a serious oversight not to thank my wife Chris, who has patiently endured my preoccupation with this project over the last three years.

One final point to emphasize is that I do not contend to establish with this book whether Alexander was right to weep. The modern version of the multiverse is not established science. In the past, proponents of multiple worlds have repeatedly turned out to be right. But that doesn't necessarily mean today's multiverse proposals will turn out to be right as well. Whether a multiverse exists, nobody can say for sure. As my friend the cosmologist Rocky Kolb replied when I posed that question to him, "An honest answer would be that we don't yet know enough to answer this question. Scientists are allowed to say we don't know."

Many serious scientists contend, though, that it is unscientific to even consider the possibility of a multiverse. But I believe the history of the issue strongly contradicts that view. A multiverse may exist, or not, but one message from this book is that the answer to that question will come from continued scientific investigation, not by denying that the question is scientific to begin with. The question of multiple worlds has always been scientific, as its history demonstrates.

Tom Siegfried
Springfield, Virginia
May 2019

The Number of the Heavens

Introduction

The idea that multiple domains may exist takes the
Copernican revolution to its ultimate limit—even our
universe may not be the center of the Universe.

—V. AGRAWAL, S. M. BARR,
JOHN F. DONOGHUE, and D. SECKEL

N THE YEAR 1277, the bishop of Paris loosened Aristotle's grip on the science of the universe.

With the blessing of the pope and the advice of an all-star team of theologians, Bishop Étienne Tempier condemned 219 philosophical propositions, most based on the teachings of Aristotle. No longer were Parisian scholars permitted to profess that "celestial bodies have eternity of motion but not eternity of substance" (Article 93). Or that "the immediate effective cause of all forms is an orb" (Article 106).

Of special note for philosophers interested in the universe was Article 34. It banned scholars from asserting that the prime mover—God—could not make multiple worlds. Until then most philosophers had taught, following Aristotle, that the existence of more than one universe was logically inconceivable. By declaring such teachings punishable by excommunication, Tempier freed scholars in Paris and elsewhere to profess that God could create as many universes as he darn well pleased.

A century ago the physicist-philosopher-historian Pierre Duhem called Tempier's proclamation "the birth certificate of modern physics." Today most historians consider Duhem's assessment much too extreme. But even if modern physics remained embryonic, Tempier's edict definitely removed the Aristotelian barrier to exploring the possibility of a plurality of worlds. Or in modern terms, a multiverse.

After 1277 speculation about multiple universes became a common pastime for medieval philosophers. Some scholars argued that other Earths, anchoring other worlds ("world" being synonymous with universe), might be just as real as the world humans knew firsthand. Others insisted that even though God might have the power to make many universes, he wouldn't have bothered. For centuries the argument over the plurality of worlds persisted.

In fact, it never really ended.

Though supposedly buried many times, the issue of multiple universes refuses to stay dead. It has been revived repeatedly as scientists have extended their grasp on reality and expanded their view of the cosmos. In the late twentieth century, the debate flared into renewed prominence, with intellectual combatants on each side just as passionate as their medieval predecessors.

Until the 1980s most twentieth-century experts had agreed that the entirety of reality was one vast (and expanding) arena of space and time. But many now suspect that this universe is not alone. Rather it may be one of an immense ensemble, a giga-gaggle of cosmic bubbles bigger than the visible cosmos, dwarfing it to a greater extent than an ocean dwarfs a molecule of water. "What we've all along been calling the universe," says cosmologist Paul Davies, may be "just an infinitesimal fragment in a much larger, more elaborate system for which want of a better word we call the multiverse."[1]

A common reaction to this notion of multiverse is that the universe is by definition all that exists, and therefore there can be no more than one of them. But this objection misconstrues the issue.

The question is whether the universe *as currently conceived* is the whole story, or merely one book in an enormous cosmic library. "Multiverse" is a provisional label for the hypothetical froth of cosmic bubbles beyond the one we observe. Someday "universe" might be redefined to encompass what's now described as the multiverse.

It wouldn't be the first time scientists needed to redefine the universe. In fact, that's a central point in the story I'm about to tell. Definitions change. Each time scientists or their philosopher predecessors have agreed on a definition of the universe, somebody soon asked whether that universe was the whole (or the only) enchilada. Each time the answer eventually turned out to be "no." Today's multiverse debate is like a remake of an old movie, just with better technology.

If that movie began with Bishop Tempier's edict, the prequel took place in ancient Greece. While the recurring philosophical debate over a plurality of worlds originated in the Middle Ages, it was rooted in a dispute between ancient Greeks favoring the existence of atoms and the atom deniers, notably Aristotle. Atomists contended that space was commodious enough to accommodate a countless number of worlds. But Aristotle constructed multiple logical objections to the atomists, and his philosophy prevailed. When early medieval European scholars embraced Aristotle's philosophy, a single cosmos was part of the package. One universe, finite but everlasting, was all that Aristotle would allow.

That universe, the Greeks taught, was a well-ordered cosmos, the world that had congealed from primordial chaos. It consisted of a set of nested spheres, with the Earth in the middle, the stars embedded in the outermost sphere, and the sun and planets attached to spheres in between. When medieval philosophers debated the plurality of worlds, the world they had in mind was this Earth-centered set of spheres, just as Aristotle had defined it.

But in the sixteenth century, Copernicus flipped the Aristotelian cosmos from Earth-centered to sun-centered. And so the terms of the debate—and the definition of universe—changed. Earth was no longer the center of *the* world; it became just one of several worlds orbiting the sun. The Copernican universe was what we now call the solar system—a system of planets orbiting the sun, surrounded at great distance by a sphere carrying the "fixed" stars.

Other astronomy-minded philosophers began to suspect that the fixed stars must be suns themselves. Soon the universe was redefined again. Instead of a sun-centered system of planets, it became a gigantic agglomeration of suns. In the eighteenth century, Immanuel Kant and others proposed that the multitude of stellar worlds composed an even larger system, a universe of stars identified with the Milky Way (or the galaxy, from the Greek root *gála* meaning "milk"). Its milky haze was the optical effect of numerous stars gathered into a huge lens-shaped system in which the sun was embedded. "Universe" became synonymous with "galaxy." And once again scientists wondered whether there might be more than one.

Kant and others noted that fuzzy patches of light (called nebulae) on the galaxy's outskirts might be entire stellar conglomerates like the Milky Way—"island universes" so distant that telescopes of the day could not distinguish their individual stars. For a century or so, expert opinion on the existence of island universes vacillated between skepticism and enthusiasm. But by the end of the nineteenth century, most astronomers declared the island universe idea to be dead. The entire universe seemed to be merely the Milky Way galaxy, home to the Earth, the sun, and a couple hundred billion additional stars.

But in 1924 Edwin Hubble demonstrated that those fuzzy patches were indeed island universes, galaxies as grand as, or grander than, the Milky Way. Instead of one lone galaxy embedded in the void, the universe comprised billions of such galaxies scattered through an immense (and, it was soon to be discovered, expanding) expanse of space. Humankind's conception of the universe—the totality of existence—magnified itself, almost overnight, by a factor of hundreds of billions. The single-galaxy universe became a multiverse, and so once again "universe" was redefined, to encompass all of the islands.

That universe, as understood today, is a vast realm of space that burst into existence 13.8 billion years ago under circumstances that astronomers sometimes compare to a fiery explosion, the Big Bang.[2] Since the bang, space has been expanding like the mother of all balloons. Today the part we can see stretches perhaps 80 billion light-years across. We can't see all of this universe, though; light hasn't had time to reach us from its most distant parts, and it probably never will. In fact, the cosmos may be infinite in extent. Still, it's supposedly all one space, all a single universe. All there is.

But maybe not. By the end of the twentieth century scientists had revived the multiverse in multiple forms. Efforts to explain features of the known universe produced a theory called inflation, which suggests the likelihood that multiple big bangs created numerous other expanding bubbles of space like (or, more probably, unlike) ours. Attempts to understand the mysteries of quantum mechanics led to the "Many Worlds" interpretation, which suggests that the universe constantly splits into multiple branches of events. Theories of matter's basic particles and forces implied the existence of extra dimensions of space, in which parallel universes might float like multidimensional soap bubbles.

This newfound fascination with a multiplex cosmos mirrors the medieval multiverse debate in a number of ways. Sure, science today possesses a much more accurate and comprehensive understanding of the universe than did the philosophers at the Middle Ages' young universities. Yet while the scientific details differ, in both cases scholars grappled with the same fundamental question about the architecture of existence. In both cases the prospect of multiple worlds faced vigorous opposition. In both cases the disputants drew on similar arguments.

There are, however, major differences between the medieval and modern multiverse debates. For medieval philosophers, the existence of a multiverse was a religious question, inseparable from the meaning of God's omnipotence. Today the multiverse issue is purely scientific. Those investigating the multiverse are posing deep questions about the nature of existence, not about a deity's power. The current multiverse debate also differs from previous versions in another important respect: the other universes discussed today are utterly beyond the reach of any conceivable telescope. Planets, stars, and galaxies, all once considered other worlds, revealed themselves to astronomers directly. Philosophers, scientists, and poets, too, could imagine visiting these other worlds and conversing with their inhabitants. But no such communication with today's multiple universes is physically possible. They exist (if they do exist) not only where no one has gone but also where no one can ever go. And where no signal can ever be sent.

So why do multiple universes matter? Because the existence of a multiverse may be essential for understanding the universe we live in. Scientists have long hoped that understanding the laws of nature would enable them to explain why our universe is the way it is. But stubborn questions have resisted this brute-scientific-force approach. It may very well be that the multiverse offers the key to unlocking mysteries about the cosmos that have so far baffled science's most assiduous detectives. "Even if the multiverse is itself unobservable," cosmologist Sean Carroll writes, "its existence may well change how we account for features of the universe we *can* observe."[3]

What's more, the possibility of a multiverse implies larger lessons about how science should be done. Whether a multiverse exists is relevant to what sorts of questions are sensible to ask, for instance, and to what kinds of rules the scientific process should observe. The rules of science that work well for explaining nature in a lone universe may not apply in a multiverse.

There's no greater story in science than the human quest to comprehend the cosmos. It began, in the scientific sense, with the ancient Greeks and was a central preoccupation of medieval scholars. In its modern form, this quest is clearly more sophisticated. But many of the deepest questions about existence are still the same, their answers remaining perpetually elusive. Ultimately the message of the multiverse debate may turn out to be that modern science is still, in some respects, medieval.

Out of Chaos, a Multiverse

We shall find sufficient reason to conclude, that the visible creation
... is but an inconsiderable part of the whole. Many other and various
orders of things unknown to, and inconceivable by us, may, and
probably do exist, in the unlimited regions of space.

—DAVID RITTENHOUSE, 1775

I N THE REALM OF COSMOLOGY—the science of the universe—truly shocking discoveries do not occur much more often than once in a lifetime. For most physicists of the current generation, that shock struck in 1998.

Sure, some discoveries since then have been dramatic and momentous. In 2016, for instance, physicists celebrated the announcement that gravitational waves had been detected, heralding a new era in observational astronomy. But such waves had been predicted by Einstein's general theory of relativity, and their existence had already been indirectly verified. Nobody was deeply astonished. In 2012 physicists found the Higgs boson in the Large Hadron Collider's proton-collision debris. It was the most important subatomic particle discovery in nearly two decades. But, again, the Higgs had been forecast by theorists in the 1960s. Most experts would have been much more seriously surprised if it had refused to show up.

In 1998, though, two teams of observers stunned the world's cosmologists with unexpected findings about the expansion of the universe. At the time, almost everybody thought the expansion of the universe was decelerating. But it turned out that the universe was not slowing down at all. In fact, it was expanding faster and faster with each passing second. "We were thrown a curve ball," cosmologist Michael S. Turner said at the time.[1]

Not since the 1930s, when astronomers first realized that the universe was expanding, had their description of the cosmos received such a drastic makeover. Even physicists who had proposed the possibility of an accelerating universe expressed amazement. "Yes, we anticipated it," one physicist told me. "But I never believed it."[2]

To be sure, it took a while for the impact to sink in. When the discovery was first reported, at an astronomy meeting in January 1998, news accounts emphasized the answer to a different question: How much did the universe weigh? Astronomers had long debated whether the amount of mass in the universe was sufficient to someday stop its expansion altogether. Perhaps, given enough matter, gravity could outmuscle cosmic expansion and collapse the universe (and everything in it) into nothingness—a "big crunch." But substantial evidence suggested that the universe simply didn't possess enough matter to do that. Instead space seemed destined to expand forever, albeit at an ever-decreasing rate. Measuring how much the expansion was slowing would indirectly indicate how much mass was tugging space inward. Above a critical value for the cosmic mass density, the future was in for a big crunch. Below that critical value, the universe would go on expanding forever.

When Saul Perlmutter, leader of one of the teams investigating the question, presented his results at the 1998 astronomy meeting, he declared that the verdict was in. No need to worry about any crunching. It seemed that the universe simply didn't have the heft to reverse its expansion and start shrinking. "It sure looks like we're in this regime where the universe will expand forever," Perlmutter said.[3]

Most news reports from the meeting focused on the "expanding forever" theme. (In my own report, I did mention briefly that the universe "may be expanding faster than ever," without commenting further.) But a couple of weeks later, after a small conference where the issue was discussed in more detail, James Glanz of *Science* magazine reported the more dramatic conclusion: the expansion of the universe appeared to be accelerating.

A lightweight universe did not have to accelerate. It could grow forever without expanding any faster than it does now. But the data showed clear hints of an accelerating expansion. In the following months, cosmologists around the world dropped other concerns and focused on trying to figure out what was powering the cosmic speedup. Attempts to answer that question quickly led to renewed interest in an even more profound implication: the prospect of a multiplicity of universes.

A UNIVERSAL CONSTANT

Foremost among those celebrating cosmic acceleration were advocates of an old idea of Albert Einstein's known as the cosmological constant.

In 1917 Einstein was playing around with the equations of his brand-new general theory of relativity. It was a theory of gravity, based on the combination of space and time into a unified "spacetime" implied by his special theory of relativity, published in 1905. General relativity was able to account for the subtle cases where matter didn't quite obey Isaac Newton's gravity law. Einstein's gravity was not a mutual tug of masses on each other but instead a consequence of mass's effect on spacetime itself. Mass distorts spacetime's geometry, Einstein asserted—textbook geometry describing straight lines and angles does not apply in space warped by matter's presence. Spacetime distortions produced by mass influence the motion of other masses, making it appear that all bodies attract each other. (Putting a bowling ball in the middle of a soft mattress depresses it; because of the depression, a marble on the edge of the mattress will roll toward the bowling ball.) As the legendary physicist John Archibald Wheeler masterfully summarized it, "Mass grips spacetime, telling it how to curve; spacetime grips mass, telling it how to move."[4]

Wheeler's summary is basically a succinct prose version of Einstein's key general relativity equation: spacetime geometry, represented by the symbols on the left side of the equation, is determined by the density of mass-energy, depicted on the right side. Einstein realized that since spacetime and mass-energy account for basically everything, his equation ought to describe the entire cosmos.

He hit a little snag when he noticed that a universe obeying his math could not remain static: it would grow or shrink. Yet for all anybody knew in his day, the universe was a permanent, everlasting receptacle for reality with no dynamic personality of its own. Einstein noted that the "fixed stars" move very slowly with respect to the speed of light (otherwise they wouldn't be "fixed"), so for practical purposes they could be regarded as being permanently at rest in a properly chosen reference frame. But his equations implied cosmic volatility. To preserve a stable universe, he added a new term, designated by the Greek letter lambda, to the left-hand side of his basic general relativity equation.[5]

Lambda symbolized what came to be called the cosmological constant. It represented a "universal constant, at present unknown." The revised equation, "with lambda sufficiently small," is "compatible with the facts of experience derived from the solar system," Einstein wrote. As long as this new term's magnitude was small enough, it wouldn't mess up the theory's predictions for planetary motions in the solar system. Lambda was necessary, Einstein said, "only for the purpose of making possible a quasi-static distribution of matter, as required by the fact of the small velocities of the stars."[6]

In his 1917 paper, Einstein did not explain lambda's physical meaning. In a paper the next year, though, he suggested that lambda represented a negative mass density—it played "the role of gravitating negative masses which are distributed all over the interstellar space." Negative mass would exert an influence opposite to the inward pull of gravity, preventing all the matter in Einstein's universe from collapsing. It was Einstein's way of keeping the universe a static and safe bubble of spacetime.[7]

But a few years later, Edwin Hubble burst Einstein's bubble. In 1929 Hubble showed that the universe is not static at all but rather expanding. Others—such as Willem de Sitter, Aleksandr Friedmann, and Georges Lemaître—had already developed interpretations of Einstein's math suggesting that possibility. Nobody paid much attention, though, and Einstein himself dismissed those efforts rather rudely. (To Lemaître, he said, your math is OK, but your physics is abominable.[8]) Hubble, however, built a stronger case based on observations of distant galaxies compiled largely by Vesto Slipher (1875–1969). Hubble's analysis showed that the greater the distance between two galaxies, the faster they appear to be receding from one another. Hubble was initially resistant to proclaim the logical deduction: that space itself is growing. But others quickly realized that the universe is not static after all—it is expanding. (Hubble eventually allowed the possibility of expansion, though he remained skeptical.)

You've no doubt encountered the famous balloon analogy: as you blow up a spotted balloon, the spots grow farther apart—not because they are moving but because the amount of balloon surface separating them enlarges. That's the picture that Hubble's discovery inspired. Groups of galaxies fly apart not from their own intrinsic need for speed (through space), but because the space between them stretches. So the universe seemed capable of avoiding collapse without any help from Einstein. He concluded that his cosmological constant was therefore not needed and considered his addition of lambda to his equation a great blunder.

Hubble's discovery became the linchpin in one of the twentieth century's grandest achievements: the explanation of the origin of the universe. That explanation took the form of what the astronomer Fred Hoyle dubbed (as a derogation) the Big Bang theory. Building on Einstein's equations and Hubble's work (incorporating further observations by Milton Humason), cosmologists outlined a scenario of cosmic history beginning with a fiery explosive event initiating the expansion of spacetime billions of years ago.

Small seeds of matter in that expanding space eventually grew into star-forming factories that built gigantic galaxies, each typically containing hundreds of billions of stars. Today's universe is in turn home to hundreds of billions of such galaxies, which congregate in huge clusters, which themselves form superclusters—vast structures decorating the seemingly endless expanse of space. In the decades following Hubble's discovery, the Big Bang became the dominant theoretical framework for explaining how this universe came to be the way it is today.

A SPECTACULAR REALIZATION

Despite the Big Bang theory's success and popularity, many cosmologists were uneasy about it. It left some crucial properties of the cosmos unexplained.

For one thing, on large scales the universe looked pretty much the same no matter what direction you looked. That implied that the baby universe had been thoroughly mixed—otherwise it could not have grown old so homogeneously. But such an initial mixture seemed implausible. Simple calculations showed that the visible universe was too large for all its parts to have been in close contact in the beginning. No mechanism could have mixed everything up unless it could transmit action faster than the speed of light—not possible, as Einstein had demonstrated. So the Big Bang theory just had to assume that everything started out smoothly mixed. That was not a very satisfying solution to what cosmologists called the horizon problem, so labeled because matter must have been mixed in the early universe beyond the reach, or horizon, of light-speed influences.

Another curiosity, called the flatness problem, also perplexed Big Bang experts. As Einstein had explained it, when massive bodies warped spacetime they created local deviations from flatness. But spacetime on the whole could still be flat on average, just as the surface of the Earth is on average a smooth curve, despite the deviations of a mountain range here and there. Whether spacetime on cosmic scales is flat or curved depends on the total matter (plus energy) density of the universe. Astronomers designate the measure of that density by the Greek letter omega. Omega greater than 1 signifies positive curvature (like the surface of a sphere). Less than 1 means negative curvature (like the surface of a saddle). Omega equal to 1, where spacetime is flat on average, is the "just right" matter density, the critical density separating a future big crunch from eternal expansion.

Even without a precise measurement, astronomers knew that whatever the value of omega turned out to be, it was obviously not a whole lot bigger, or a whole lot smaller, than 1. Much bigger than 1, and the universe would have crunched long ago. Much smaller and it would have expanded too quickly for stars or galaxies to form, meaning no place for planets and no home for people, even cosmologists. So it seemed rather lucky that omega was in the right ballpark for life to exist.

And it was suspiciously lucky. During billions of years of expansion, the density of matter in the universe would have constantly diminished. If omega is pretty close to 1 today—and it is—then way back at one second after the Big Bang it must have been much more extremely close to 1: no greater than 1.000000000000001 or less than 0.999999999999999. For no obvious reason, the universe began with a mass density unbelievably close to omega equal to 1; had it been only slightly different, today's universe would look nothing like it now appears. In other words, the universe is flatter than it has any right to be. That's the flatness problem.

As so often happens in physics, the solution to the horizon and flatness puzzles did not come from efforts to solve those problems.

It came out of the blue, from an attempt to solve a completely different, and apparently unrelated, problem, involving a quandary about making half a magnet.

Back in the days when magnets were popular toys (rather than refrigerator decorations), many children discovered an amazing fact about the physics of magnetism. If you break a bar magnet in two, you double the number of magnets you have. Each bar retains two poles (usually designated north and south). No matter what you do, you cannot create a magnet with only one pole.

In theory, though, single-poled magnets could exist in the form of an exotic subatomic particle known as a magnetic monopole. Theoretical calculations suggested that the universe should be full of magnetic monopoles, created in the cauldron of the superhot Big Bang. In fact, monopoles ought to be as common as protons, and much more massive. Yet no sign of one has ever been seen. (Blas Cabrera of Stanford University did report a monopole sighting in 1983, but that report was not confirmed and was retracted in 1990.)

Among those investigating the monopole problem in the late 1970s was an unassuming and affable young physicist named Alan Guth. Guth had long been interested in monopoles, and his studies eventually led him to contemplate their connection to the hottest topic in physics at the time—the effort to unify theories of nature's fundamental forces: electromagnetism and the strong and weak nuclear forces. (Gravity, the fourth fundamental force, stubbornly resisted efforts to include it in the package.) Electromagnetism and the weak nuclear force had already been successfully united in the 1960s. Adding the strong nuclear force produced what came to be called grand unified theories, or GUTs. Guth preferred to call them GUThs.

Collaborating with Henry Tye of Cornell University, Guth had shown that GUTs, if correct, did indeed forecast the production of multiple monopoles. Similar work by John Preskill at Caltech rigorously established that there was indeed a monopole problem: too

Alan Guth

many predicted, too few (as in zero) detected. While at the Stanford Linear Accelerator Center in 1979, Guth pondered the problem further. Perhaps, he conjectured, the discrepancy involved the details of the GUTs. He worked through the physics that GUTs implied for the early universe and reached what he described in his notebook as a spectacular realization: the standard accounts of the Big Bang theory had left out a very important chapter. He called that chapter "inflation."

FALSE VACUUM

Guth's spectacular realization involved what physicists refer to as a phase transition, a shift from one state, or phase, to another. A phase transition is what happens when liquid water turns to ice, or vice versa. It can happen gradually (like a block of ice melting in the sun) or almost instantaneously. A famous but fictional example of a rapid phase transition appears in a story by Italian journalist Curzio Malaparte, who spent a winter in Russia during World War II. In Malaparte's story 1,000 horses rushed into Lake Lagoda to escape a forest fire. After the horses entered the lake, the water suddenly froze, trapping them almost instantly.[9] While this incident never really happened, the principle behind the physics of it is well known. Water in a sufficiently pure state can get very cold, far colder than the normal freezing point of zero degrees Celsius. Water in those conditions is said to be supercooled; it appears stable, but it's a "false" stability, owing strictly to the absence of any disturbance. The transition to ice requires a trigger. A speck of dust dropped into such water (let alone a horse) can induce it to suddenly freeze.

Guth envisioned a similar phase transition in the early universe. As the Big Bang fireball cooled, cosmologists presumed, various forces described by grand unified theories would appear. At first, when very hot, the forces would seem as one, like when a collection of H_2O molecules is hot enough to exist completely as steam. As steam cools, though, some liquid drops will form and eventually ice crystals too—new states of matter generated in phase transitions. Similarly the cooling universe was marked by phase transitions that generated the strong nuclear force, weak nuclear force, and eventually electromagnetism.

But suppose that, like the falsely stable supercooled water, the universe itself existed in a state of "false" vacuum. In that case the background energy would not be in a truly stable state. If so, Guth calculated, at a tiny amount of time after the Big Bang (something like a trillionth of a trillionth of a trillionth of a second) space essentially went "poof!"—the volume of the universe doubling over and over, several dozen times. At that rate, an expanse of space-time the radius of a proton would grow almost instantly to the size of the planet Jupiter. It seemed to Guth that "inflation" aptly described this vast and sudden expansion of space.

He quickly saw that inflation explained the monopole problem. Monopoles very well may have been produced in the very early universe. But if space had expanded by such a huge factor after the monopoles were created, their concentration would have been drastically diluted. Chances were slim that any would be around in our corner of the universe today. The absence of monopoles suggested that Guth's false vacuum–inflation scenario had in fact occurred.

Guth also realized that his scenario solved the flatness problem. Inflation would have ballooned the universe so drastically that any initial curvature would have been virtually eliminated, no matter how curved spacetime was to begin with. Imagine holding a croquet ball in your hand—its curvature is obvious. But if it instantly expanded into a sphere the size of the Earth, you could stand on it, and it would look perfectly flat. A greatly inflated space would similarly seem very nearly flat as well.

Soon Guth's new theory added a third trophy to its résumé, when he learned about the horizon problem. What we see in the universe today could have been all mixed up in the beginning after all, because inflation began within such a tiny patch of space. Before inflation, the radius of the currently visible universe might have been about 10^{-52} meters. After inflation, the universe would have been about one meter across, an enormous increase in size. (Keep in mind that these estimates apply to the size of the observable universe—the region of spacetime that is today accessible to our telescopes. The whole universe beyond could be vastly larger still, even infinitely large, as current evidence suggests.[10]) Nobel laureate Murray Gell-Mann, famously the most supercritical physicist of his era, grasped the implications of Guth's explanation for the horizon problem immediately. "You've solved the most important problem in cosmology," Gell-Mann told him.[11]

Guth began lecturing on his new idea around the United States and Europe. As he well knew, though, his theory had its flaws. It may have solved some of cosmology's big problems, but it had internal issues of its own, such as figuring out how inflation would have stopped. Guth's theory explained how the universe could have rapidly inflated but did not explain why it had soon settled into a nice leisurely expansion. At least in our realm of reality, inflation must have ended soon after it started, otherwise our spacetime bubble would have been blown apart so rapidly that no stars and therefore no galaxies would have been able to form.

Attempts to solve what came to be known as the "graceful exit" problem soon appeared, based on modifications of Guth's original idea. Other issues arose, though, such as whether an inflationary scenario could produce the arrangement of galaxies in clusters and superclusters observed throughout the cosmos. During the 1980s and 1990s, physicists addressed these issues by developing a zoo of inflationary variants, with names such as double inflation, soft inflation, extended inflation, and hyperextended inflation. But all retained the basic element of rapid early expansion making the universe homogeneous and flat. And at least some of the variant versions of the theory suggested an additional implication of inflation: the likely existence of other universes.

INFLATIONARY CHAOS

In 1982 physicists gathered in Cambridge, England, for the Nuffield Workshop, a conference where they discussed the latest developments in cosmology. Among the attendees was Andrei Linde, a young physicist from Moscow. Linde is widely known as one of cosmology's most colorful characters; he combines a witty sense

Andrei Linde

of humor with acute condemnations of viewpoints or presuppositions that he finds illogical or erroneous.

Linde had quickly realized that Guth's inflation, while ingenious, needed work. One problem involved the need to produce matter in the infant universe. In Guth's theory, inflation would have produced many bubbles in spacetime during the universe's early phase transitions. Supposedly the bubbles would eventually all merge, getting the entire universe out of the false vacuum. When bubble walls collided, energy would have been released and converted into matter, providing the baby universe with a hot soup of elementary particles. Sadly for this scenario, though, further calculations by Guth and his colleague Erick Weinberg showed that the bubbles would not necessarily merge. Without those mergers, no matter would be produced inside the bubbles. Most of the energy would remain trapped in bubble walls. Guth and Weinberg considered the possibility that one big bubble could have inflated to create today's universe but found that such a bubble also would be empty of matter.

Linde, though, examined other scenarios and developed a new version of inflation (creatively described as "new inflation") that did allow matter to form in a single bubble. (A similar idea was developed independently in the United States by Paul Steinhardt and Andreas Albrecht.) During the Nuffield conference, Linde articulated an important ramification of new inflation: if it happened once, it would have happened many times. Inflation not only produced a universe, it produced a multiverse.

Linde's new inflation allowed the bubbles to remain independent, while particles of matter could be created within a bubble by oscillations of an energy field. This field did not need to belong to one of the fundamental forces, though. Linde viewed it as an independent field that came to be called the "inflaton." This process of creating particles through energy-field oscillations, referred to as thermalization, could produce a matter-containing universe like the one we see. But other bubbles that we can't see would also exist. "Each bubble can be considered as a mini-universe which has no possibility of any causal contact (e.g. by collision) with other mini-universes," Linde declared. "Each mini-universe (each bubble) has its beginning in time . . . and its evolution after thermalization can be described by the usual hot universe theory."[12]

As it turned out, even the new inflation had its problems. Soon Linde developed a still-newer version, called chaotic inflation, to address new inflation's technical challenges. Linde's chaotic version retained the multiplicity of spacetime bubbles. Within those bubbles he foresaw the solution to another problem that had been nagging some cosmologists: the universe's suitability for life. Certain features of physics—such as the values of fundamental constants governing the strength of various forces, the masses of basic particles, and the expansion rate of the universe—all seemed precisely tuned to allow life to exist. Had these basic properties differed from their known values, life's biochemical complexity could never have been realized. In his paper written at the Nuffield conference, Linde argued that inflation's multiplicity of bubbles could explain the universe's suitability for life—by invoking what was known as the "anthropic principle."

"In the scenario suggested above the universe contains an infinite number of mini-universes (bubbles) of different sizes, and in each of these universes the masses of particles, coupling constants, etc. may be different due to the possibility of different symmetry breaking patterns inside different bubbles," Linde wrote. "This may give us a possible basis for some kind of Weak Anthropic Principle: There is an infinite number of causally unconnected mini-universes inside our universe, and life exists only in sufficiently suitable ones."[13] In other words, a multiverse provided a logical way to explain the anthropic suitability of our universe's properties.

Most physicists then (and even now) did not like the anthropic principle very much (as you'll see in chapter 12). Only Linde and a handful of others pursued the notion of a multiverse as a way of explaining the habitability of the cosmos. "The first ten years of the journey through the multiverse were very exciting but rather lonely," Linde recalled years later. "The standard lore was that one should avoid using anthropic arguments for explaining fundamental properties of our world. As a result, some of the key ideas relating to each other inflation and anthropic considerations originally were expressed in a rather cryptic form."[14]

During the 1990s, though, the multiverse's popularity began to rise. One factor noted by Linde was the appearance online in 1991 of the arXiv.org preprint server, originally hosted at the Los Alamos National Laboratory with the website address xxx.lanl.gov. Physicists had long shared preprints—preliminary versions of papers to be submitted for publication—among themselves. Snail mail and fax machines could distribute such papers here and there, but not very efficiently. With the establishment of arXiv.org, physicists could post papers without the delays and roadblocks of peer review for publication in paper journals. New ideas could fly around the globe instantly, available to anyone who was interested and had a connection to the Internet. Ideas about the multiverse could easily be widely disseminated, even if traditional journals—and most physicists—didn't pay much attention.

Then in 1998 that dramatic once-in-a-lifetime discovery of cosmic acceleration reignited interest in the multiverse in a big way. When physicists found that the universe's expansion rate is accelerating, they could no longer ignore the multiverse. They needed it.

SPEEDING UP

Cosmic acceleration implied that something like Einstein's cosmological constant—a residue of energy in the vacuum of space—existed after all, driving space apart. But most physicists had concluded that the cosmological constant had a value of zero (same as not existing). Calculations showed that if the cosmological constant did exist, it would have been powerful enough to blow the universe up much too rapidly for anyone to live in it. It was therefore natural to surmise that some unknown mechanism had instead canceled the cosmological constant out of the universe entirely. On the other hand, the presence of such vacuum energy might be just what cosmologists needed to complete the recipe of the universe's ingredients.

Inflation theory's success had implied that something was missing from that recipe. All the light-emitting matter observed in space as stars and luminous gas could not possibly provide enough mass to make the universe flat (corresponding to omega equal to 1), as inflation required. True, astronomers had already established that the way galaxies spin and cluster implied the existence of far more matter than could be seen. But even that "dark matter" did not seem abundant enough to make the universe flat. Some new cosmic ingredient would be needed to reconcile a flat universe with observations.

Identifying that ingredient became a hot topic in physics after the news in January 1998 about the universe's accelerating expansion. Physicists were still buzzing about it in April, when the American Physical Society held one of its two major annual meetings (this one always called the April meeting, even when it's held in May, or sometimes January) in Columbus, Ohio. So the room was packed when Robert Kirshner, a leader of the team competing with Perlmutter's, presented an overview of the methods used to measure the cosmic expansion.

Both teams had taken basically the same approach: analyzing the light from distant exploding stars—supernovas. Traveling over vast expanses of space, light from supernovas gets stretched by the Doppler effect, just as the pitch of a receding siren gets lower and lower.

With light, the frequency diminishes and the color gets redder as the expanding universe increases the distance between the light's emission and its observation. This change in color, known as redshift, can be used to infer how fast the universe is expanding—if you know how far away the supernova is.

Fortunately, one type of supernova allows for an estimate of its distance independent of its redshift. That version of a supernova, known as type 1a, occurs when a white dwarf—the Earth-sized remains of a dead ordinary star—assimilates an excessive amount of matter and explodes as if because of overeating (but actually more like a thermonuclear bomb). By basic textbook accounts, all type 1a supernovas explode when a white dwarf's mass exceeds 1.4 times the mass of the sun—the densest a planet-sized body can be without triggering a chain reaction and blowing itself up.

Presumably, since all type 1a explosions destroy the same amount of mass, they should emit the same amount of energy and therefore shine with equal brightness. So their apparent brightness seen from Earth should depend only on their distance, making them what astronomers like to call "standard candles." Determining the distance of a nearby 1a by another method would enable a calculation of the distance to those much farther away. Then, by measuring how much the light emitted from supernovas is stretched at different distances, astronomers could deduce how fast the universe was expanding at different epochs in time. In that way they could determine how the expansion rate changed over the course of cosmic history.

In practice it's not so simple. It turned out that type 1a supernovas are not really all equally bright. A type 1a's brightness might depend, for instance, on exactly how the white dwarf star acquired the extra mass needed to push it over the 1.4 solar-mass barrier, known as the Chandrasekhar limit. Textbooks said the dwarf siphoned hydrogen gas from a nearby companion star. But experts knew that that was more of a guess than well-established astrophysics. A white dwarf might also explode if it collided with another white dwarf. In any case, some type 1a supernovas can be several times as bright as others. So to use type 1a supernovas as standard candles, astronomers had to develop methods to correct for brightness differences. Luckily the color of light that type 1a supernovas emit, as well as how they dim over time, depends on how bright they were to begin with. Applying corrections for these factors, astronomers gained confidence that they could reliably estimate the true brightness, and thus the distance, to many faraway stellar explosions.

But then the results of the observations and calculations surprised everybody. Faraway supernovas seemed dimmer than they were supposed to be. In a decelerating universe, as Kirshner explained, a packet of light (photon) traveling to Earth from a supernova would have less distance to travel, and therefore would appear brighter than it would in a universe expanding at a constant rate. But if cosmic expansion accelerates, a photon has farther to travel and would show up at Earth somewhat dimmer than expected.

Kirshner, a congenial astrophysicist at Harvard (and pretty much a dead ringer for the golfer Tom Watson), had long been one of the world's top experts on supernovas. I had consulted him frequently in writing about them, including the famous supernova 1987a in the Large Magellanic Cloud, a satellite galaxy to the Milky Way. At the meeting in Columbus, Kirshner explained the implications of his team's supernova measurements for cosmic expansion history. "It looks like the evidence favors a universe that has been actually speeding up," he concluded. But he emphasized that there was still some uncertainty. "It's not beyond a reasonable doubt, just a preponderance of the evidence," he said. "It's a civil case."[15]

I saw Kirshner again a couple of weeks later at a conference at Fermilab (full name, the Fermi National Accelerator Laboratory) in Illinois, an hour or so west of Chicago. There a few dozen of the world's leading cosmological thinkers gathered to ponder the supernova results. Kirshner emphasized once more that the case for acceleration was not definitive. It did seem clear that the universe did not contain enough matter to crunch. Rather than a value for omega, the cosmic energy density, equal to 1 (the crunch threshold), omega appeared to be in the range of 0.2 to 0.3.

Whether the universe was actually speeding up depended on the value of another measure, the cosmic deceleration parameter, designated q-naught and symbolized by q_0. A q-naught of 0.5 corresponded to omega equal to 1. A q-naught less than that corresponded to a universe expanding forever. But, as Kirshner explained, the supernova data implied the astounding possibility that q-naught was *negative*. In that case cosmic expansion is not decelerating at all but rather progressing faster and faster. "From our data it looks like it could be accelerating," Kirshner said.[16]

If the expansion of the universe is accelerating, though, there must be more within it than just ordinary matter and energy. Matter exerts gravitational attraction, pulling space inward. Energy, in the form of radiation such as light, exerts pressure. And Einstein's general theory of relativity showed that this pressure also contributes to the inward gravitational force. So acceleration requires some additional component of space, other than matter and radiation, to drive space apart. Such a component would need to exert "negative pressure," causing space to expand.

Many of the talks at the Fermilab conference focused on what the component of space driving the acceleration might be. First on the program was Michael Turner. He offered a possible answer. Space, he said, must be filled with what he called "funny energy." "Current observations, taken at face value, tell us that most of the universe is funny energy whose pressure is negative," Turner said.[17]

The right amount of this funny energy would resolve the mystery of omega. It's not just matter that contributes to omega, but the universe's total mass-energy density. An energy component added to the mix could make omega equal to 1, just as inflation ordered. In fact, later observations suggested that the combined recipe of ordinary matter, dark matter, and funny energy did add up to very close to 1—the matter making up 25 to 30 percent of the cosmic mass-energy, funny energy making up 70 to 75 percent.

So inflation fans could celebrate: the missing ingredient in the cosmic recipe had been found. But on the other hand, they didn't really know what it was. Funny energy is just a name, and in any case, nobody really liked that name. Turner later rechristened it "dark energy" in analogy to dark matter, and that label stuck. But its true identity remained a secret. "We have no real clue for what this stuff is," said Fermilab theorist Josh Frieman.[18]

Still, dark energy's properties sounded familiar. A negative-pressure energy filling all of space precisely matched the features of Einstein's cosmological constant. Its negative pressure would be just the thing to drive an accelerating expansion. "It was good enough for Einstein," said Turner. "It ought to be good enough for us."[19] So perhaps Einstein's big blunder was turning out to be one of his most spectacular predictions about the universe. He had no idea, though, that his source of negative pressure driving cosmic acceleration would also drive many scientists to contemplate a multiverse.

WHY ME? WHY NOW?

Of course, inflation theory had already suggested the possibility of a multiverse. But the discovery of accelerating expansion gave the multiverse idea a boost. Without a multiverse, it seemed hard to explain why cosmic acceleration was going on now, in an epoch of cosmic time during which people were around to discover it. Turner liked to call it the Nancy Kerrigan problem: "Why me, why now?" In her case, she wondered why some thug bashed her knee right before the Winter Olympics, where she would have been a contender for a figure skating gold medal. With the universe, the question was why cosmic acceleration began relatively recently, only a few billion years ago ("recent" from a cosmic-time perspective). Dark energy of just the right "Goldilocks" strength would be needed for acceleration to begin when it did. Why should dark energy be just the right strength?

Inflation's multiplicity of universes provided a convenient answer. Its multiple bubbles could possess different amounts of dark energy; one of them had to have the right amount to resolve the Nancy Kerrigan problem. Multiple universes are not a bug in inflationary theory; they are a feature. Inflation produced not only a better understanding of the universe but also a new conception of the nature of the universe. Its explanation for how the observable universe came to be entailed within it the recipe for building many others.

It was not the first time anyone had concocted such a fertile cosmology. Something very similar had happened in the thirteenth century, in a theory of the universe's origins devised by a mental magician named Robert Grosseteste.

2

Robert Grosseteste's Multiverse

This universe is only one of an infinite number. Worlds without end.
Some benevolent and life-giving, others filled with malice and
hunger. Dark places where powers older than time lie ravenous
and waiting. Who are you in this vast multiverse, Mr. Strange?

—THE ANCIENT ONE, *DOCTOR STRANGE,* 2016

ALISTAIR CROMBIE thought medieval science didn't get enough respect.

Around the halfway point of the twentieth century, he wrote a book aimed at correcting the common misperception that the Middle Ages was a dark time for science and investigating nature. Crombie's own studies had convinced him that much more had been accomplished in the early centuries of the second millennium than other historians (and scientists) had generally recognized. The so-called "dark ages" preceding the sudden birth of modern science had in fact, Crombie maintained, been the prelude to the Age of Enlightenment. Of special importance, he believed, was the scientific philosophy and methodology developed at the University of Oxford by the brilliant polymath Robert Grosseteste. "What seems to be the first appearance of a clear understanding of the principles of modern experimental science is found in the writings of the English logician, natural philosopher, and scholar, Robert Grosseteste," Crombie wrote. "In the thirteenth century the

Oxford school, with Robert Grosseteste as its founder . . . marks the beginning of the modern tradition of experimental science."[1]

Most other historians disputed Crombie's contention. A more recent assessment by the late David Lindberg considered it seriously extreme. "No informed scholar of my acquaintance accepts Crombie's claim that Grosseteste was the founder of experimental science," Lindberg wrote.[2] Crombie himself acknowledged in later writings that he had overstated his case: "Some of the expressions I used about the extent of the medieval contributions to the structure and methods of research of modern experimental science now seem to me exaggerated."[3]

But Crombie was not wrong to say that the intellectual achievements of early medieval Europe had been neglected and that the scientific enterprise in the Middle Ages was more vigorous and sophisticated than commonly portrayed. And as Lindberg assented, Crombie was also on target in suggesting that Grosseteste was a superstar of medieval science, even if not the founder of modern methods. To those aware of him, Grosseteste is a legendary, larger-than-life character. "Truly Grosseteste was one of the great encyclopedic thinkers of the world," his biographer Francis Seymour Stevenson wrote in the nineteenth century.[4]

Grosseteste was regarded as an expert on virtually every field of knowledge. He wrote influential treatises on sound and heat, comets, rainbows, and optics. He was also an early cosmologist, proposing a scenario for creating the entire Aristotelian universe—the series of concentric spheres that carried the planets and stars around the Earth (then believed to reside at the cosmic center). In proposing a mechanism for constructing the cosmos, Grosseteste also unwittingly outlined a scheme capable of creating a multiverse. Just as late twentieth-century theorists forged an explanation of cosmic origins that implied multiple universes, Grosseteste's attempt at an origin theory for the thirteenth century's version of the universe revealed a way for making many other different universes as well.

BISHOP OF LINCOLN

As with most medieval legends, many aspects of Grosseteste's life are hard to verify.[5] It seems that he was born in Stowe, County Suffolk, sometime around 1170. His parents were poor, descendants of Normans who came to England with King William's conquest. Consequently Grosseteste's native language was French. But he no doubt learned English as well, and his scholarly writings were, of course, in Latin.

Apart from evidence for his family's poverty, no information exists about his youth and early education. He may have attended a cathedral school at Hereford. There are hints that he studied law, and also medicine. He certainly studied theology, probably in Paris. Eventually he made his way (or returned) to Oxford, where he became chancellor—by some accounts, the university's first. "He was the one really outstanding personality the university produced during the first generation of its existence," philosopher James McEvoy wrote.[6]

Little else is known of Grosseteste's early career. And much that some past biographers claimed about it is probably wrong, modern scholars have shown. But whatever the details of his educational background and accomplishments, there's no doubt that Grosseteste was a man of prodigious intellect and enormous integrity. In 1235 he became the Bishop of Lincoln, making him one of the most prominent and powerful church officials in England until his death in 1253. Much of his reputation over subsequent centuries stemmed from his role in religious issues. He was a voice against corruption and condemned church politics when they conflicted with Christian teachings. He even argued with the pope.

Grosseteste's personal integrity in dealing with affairs of church and state mirrored the intellectual integrity with which he investigated the natural world. His work as a scholar was deeply influenced by the writings of Augustine of Hippo, known to us as Saint Augustine, who in turn had assimilated much of the philosophy of

Plato. Grosseteste was also one of the first medieval scholars in western Europe to study Aristotle thoroughly. "Grosseteste . . . tended to be curious about the historical origins of ideas and the kind of men who had originally thought them up," McEvoy wrote.[7] Grosseteste was not blindly devoted to Aristotelian authority, though. Observations settled questions about nature's behavior, Grosseteste believed, not Aristotle's writings.

Grosseteste also emphasized the importance of mathematics for intellectual rigor in investigating nature. In particular he stressed the key role of geometry (lines, angles, and figures) in the study of optics, which in those days was considered a—or even *the*—fundamental science, much like particle physics or cosmology today.

He did not possess the mathematical apparatus that is now available for describing the cosmos. But he did attempt his own version of a creation cosmology, built on the notion of light as the fundamental principle of existence. In his treatise on light, he described a process for building a universe—not the universe as we conceive it today, of course, but rather the medieval conception of the universe descended from Aristotle.

THE MEDIEVAL UNIVERSE

Grosseteste sought to explain the origin of a cosmos that had first been envisioned by early Greek philosophers. They perceived a need to describe reality rationally, conceiving a "world"—a totality of existence—in which the drama of events proceeded in a regular manner rather than reflecting the chaos of divine whimsy. Those philosophers envisioned "an orderly, predictable world in which things behave according to their natures."[8] The Greeks called that world the *kosmos*. Humans have ever since been trying to understand what that *kosmos* actually is, and how to define it. And whether there is more than one.

Long before the Greek philosophers began their cosmic speculations, skywatchers in Mesopotamia and elsewhere had noticed certain patterns in the movements of the lights dotting the night sky. In their effort to comprehend their *kosmos,* the Greeks pondered those patterns and tried to infer from them the mechanism of the heavens. Several features of those cosmic motions required explaining: the nightly rotation of the stars around the sky; the appearance of the sun in different segments of the sky during different seasons; the wandering of the planets through the background of the "fixed" stars; and the curiosity of "retrograde motion"—some of the planets appeared to reverse their direction on occasion and move backward.

Before Plato, philosophers had recognized that the stars rotated around the sky as though they were points of light on the inner surface of a sphere. Eventually a consensus emerged that the Earth itself was also a sphere, and it seemed that the stars and other heavenly bodies somehow rotated around it.

Assuming the Earth to be a sphere, Greek astronomers could then define a celestial equator—the plane that passes through the Earth's equator. It was obvious that the sun did not orbit the Earth in that plane. Instead it circled at an angle. The plane of the sun's orbit—called the ecliptic—was offset from the celestial equator by an angle of 23½ degrees. (Today, of course, we know that the Earth's axis is tilted by that angle as it orbits the sun, rather than the sun orbiting the Earth at that angle.) For both ancient and medieval astronomers, the relationship between the equator and the ecliptic guided the analysis of heavenly motions.

One particularly peculiar motion complicated the cosmic picture. If you imagine the equator's polar axis—the line extending upward perpendicular to the equatorial plane and passing through the center of the Earth—you'd find that it doesn't maintain the same position with respect to the axis of the ecliptic. Rather, the polar axis for the equator seems to rotate about the ecliptic axis. Because of that rotation, the points where the ecliptic intersects the equator (marking the spring and autumnal equinoxes) change over time, returning to any given spot every 26,000 years. (The

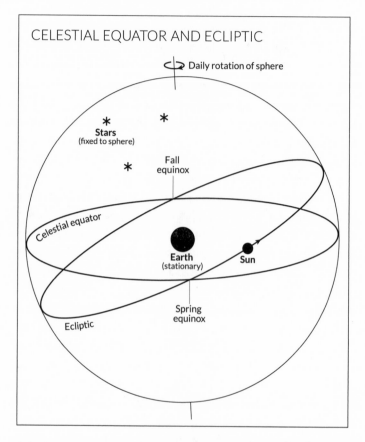

CELESTIAL EQUATOR AND ECLIPTIC

Daily rotation of sphere

Stars
(fixed to sphere)

Fall
equinox

Celestial equator

Earth
(stationary)

Sun

Spring
equinox

Ecliptic

Celestial Equator and Ecliptic. Astronomers in ancient Greece believed that the heavenly bodies revolved on spheres centered on a spherical Earth. The path of the sun around the Earth defined the ecliptic. The plane passing through the Earth's equator defined the "celestial equator." Spring and autumn began on Earth when the sun's path (ecliptic) crossed the celestial equator. Now we know that the Earth orbits the sun, and the angle offsetting the ecliptic from the equator is actually due to the tilt of the Earth's axis with respect to the plane of its orbit.

Greeks didn't have that number exactly right; they thought it was more like 36,000 years.)[9]

In Plato's time it remained unclear exactly how to account for these motions of the sun, moon, and other planets. A lover of mathematics (and worshiper of the planets as gods), Plato challenged astronomers to ascertain the mechanisms that produced the observed celestial phenomena. As circular motion was divine motion, Plato insisted that the goal be achieved by identifying some combination of uniform circular motions.

Such a mathematical system appeared in the fourth century BC from Eudoxus of Cnidus (c. 400–347 BC), who may have spent some time studying with Plato in Athens.[10] Eudoxus was an exceptional mathematician, especially skilled at geometry, as well as a scholar of geography. When he applied math to the heavens, he envisioned a set of concentric spheres, embedded one within another like a family of Russian dolls. Each celestial body was attached to one of these spheres. Collectively the spheres rotated in just the right way to reproduce the observed motions in the sky.

Eudoxus said nothing about what the spheres might be made of or how they moved and interacted physically. So historians traditionally have believed that he was making no claim at all about the actual physical construction of the universe. Plato worshiped math even more than he loved planets, and his exhortation was to find a mathematical description of celestial motions, not to devise a physical model of the cosmos. You could use the math of rotating spheres to predict such things as where the sun would rise, but that didn't necessarily mean the spheres were real. Instead the goal was "saving the phenomena"—accurately describing observations of the heavens mathematically, not physically. (This approach to science was common into the Middle Ages and beyond, and traces of it remain stuck in the minds of many scientists even today.)

Some modern scholars aren't so sure about what Eudoxus was thinking. His silence on physical matters is not proof, they argue, that he thought his system was just a computational device. To the contrary, maybe Eudoxus believed he had found a physical description based on Plato's insistence that uniform circular motion ruled the skies. After all, as a tool for saving the phenomena, the Eudoxian spheres didn't work that well. They predicted some planetary positions and motions that did not always match observations. One scholar has argued that if Eudoxus had been interested only in mathematical predictive power, he could have improved and simplified his system. But that would have been possible only at the expense of deviating from the physical requirement of uniform circular motion.[11] It is nevertheless true, though, that in ancient and medieval times, astronomy was widely regarded as a mathematical rather than physical science.

If Eudoxus did consider his system to be physically real, it was not a simple thing to visualize. It did not consist merely of one sphere per planet; a planet needed several spheres, each responsible for one or another aspect of its complicated motion. Altogether the Eudoxian cosmos required 27 spheres. A later refinement, by Callippus of Cyzicus, made the predictions of the Eudoxian system more accurate by adding seven more spheres, for a total of 34.

NATURAL MOTION

Unlike Plato, Aristotle was more a fan of physics than math. So he deliberately composed a description of the cosmos designed to turn Eudoxus's spheres from mathematical fictions (if that's what they were) into physical engines.

Earth's residence at the center of the universe was the central point in Aristotle's system. Aristotle held that the Earth was spherical and motionless. It was surrounded by spheres carrying the other planets and the fixed stars. This set of spheres constituted the entirety of existence—the world, or universe. For the Greeks, *world* and *universe* meant the same thing: the whole cosmos. Synonymous use of *world* and *universe* continued into the Middle Ages, so when philosophers spoke of "many worlds," they meant many universes—as the notion of "universe" was then conceived.

Aristotle's cosmic model was part of his larger devotion to providing a solid logical foundation for understanding matter and motion. He believed he had demonstrated his propositions about the world beyond doubt by elaborate syllogistic reasoning (aided by key observational evidence). His logical proof of the Earth's central location rested largely on his notion of "natural motion"—different kinds of matter moved naturally toward their proper place in the universe.

In those days the "kinds" of matter referred to the four "elements": fire, air, water and earth. Unless impelled otherwise by some constraint, those elements moved naturally in straight lines: earth and water downward (toward the center of the universe), air and fire upward (away from the center). That explained the Earth's location—it occupied the center of the cosmos because that's where earth's and water's natural place is.

Yet while air and fire moved upward toward the heavens, the heavens themselves appeared to be composed of something else. Natural motion of the four elements was "simple" motion, always a straight line, either toward or away from the cosmic center. (Unnatural, or "violent," motion occurred because of the influence of some other action, such as a person throwing a rock.) But heavenly motion appeared to be circular—the only other kind of simple motion.

"Our eyes tell us that the heavens revolve in a circle," Aristotle wrote in his treatise *De Caelo,* or *On the Heavens.*[12] That circular motion was everlasting and changeless: "In the whole range of time past, so far as our inherited records reach, no change appears to have taken place either in the whole scheme of the outermost heaven or in any of its proper parts."[13]

Such circular motion, while unnatural for the four known elements, was obviously natural for the heavens, Aristotle argued. And if the natural motion of heavenly substance was circular, it must therefore have been some fifth kind of substance, a different element. "We may infer with confidence," Aristotle wrote, "that there is something beyond the bodies that are about us on this earth, different and separate from them."[14] Aristotle therefore conceived the cosmos as two realms: the messy, ever-changing earthly sphere of the four known elements, and the perfect, changeless ethereal realm of the heavens, built by the fifth element, commonly referred to as ether.

(It's just a coincidence, of course, that modern-day astrophysicists have also concluded that celestial substance—the bulk of the matter throughout the cosmos—is some form unknown on Earth, referred to simply as "dark matter." It is nevertheless intriguing that the ancient question of cosmic composition—what the universe is made of—remains unanswered today.)

In Aristotle's cosmos the "planets"—moon, sun, Mercury, Venus, Mars, Jupiter, and Saturn—each occupied a sphere that revolved around the central Earth. Aristotle did not spend much time worrying about the order of those spheres. Other philosophers proposed planetary orders that did not agree with one another. Claudius Ptolemy (c. AD 100–170) initially positioned them in the order moon, Mercury, Venus, sun, Mars, Jupiter, Saturn. But there was still much debate, often focused on whether Mercury and Venus were below the sun or above it. In later works Ptolemy even changed his mind about the position of the sun and decided that the proper order could not be ascertained for sure.[15] Everybody agreed, though, that the stars occupied a sphere above all the planets.

In explaining how the heavens worked, Aristotle invoked the spheres described by Eudoxus. But Aristotle clearly desired more than just working out the math that predicted the motions in the sky—saving the phenomena. He sought a physical description. He realized that to make a real moving system of physical spheres actually work required more complexity than you need just to make mathematical forecasts. His physical depiction of the cosmos required many intermediary spheres to modify each primary sphere's

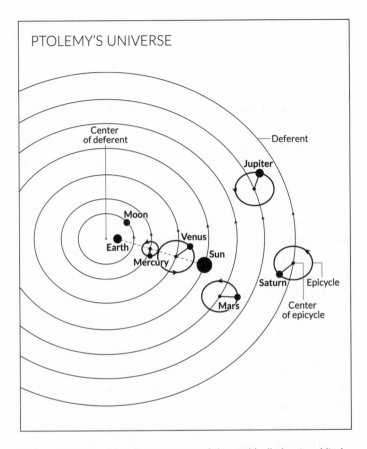

PTOLEMY'S UNIVERSE

Center of deferent

Deferent

Jupiter

Moon

Venus

Earth

Sun

Mercury

Saturn

Epicycle

Mars

Center of epicycle

Ptolemy's Universe. In Ptolemy's system of the world, all planets orbited a central Earth, but not in simple circles. A planet actually turned around a smaller circle, called an epicycle, centered on a point that traveled along a circle around the Earth. Even those circles (called deferents) had centers that were offset from the actual position of the Earth.

motion (and to keep the motion of the outer spheres from messing with the inner ones). Every planetary sphere required several additional spheres, yielding a total of 47. Or 49. Or 55 or 56 or 61. (There is some ambiguity in Aristotle's descriptions, so scholars don't all agree on exactly how many spheres he required.[16])

Still, Aristotle's cosmos didn't really work all that well. All his spheres rotated about a point occupying the cosmic center (occupied by the Earth), and therefore the planets attached to those spheres should have maintained a constant distance from Earth. Yet planets varied in brightness from time to time, suggesting that their distance must change. And while some motions, such as the retrograde reversal of planetary directions, could be explained qualitatively with rotating spheres, details of such motion eluded precise prediction.

These defects were largely corrected a few centuries later by Ptolemy, the greatest astronomer of ancient times (and, like Eudoxus, a skilled mathematician and geographer). Ptolemy's mathematical approach described circular orbits modified by epicycles—circles around circles, like the way (we now know) that the moon orbits the Earth as it orbits the sun. Ptolemy's system also required other geometrical sleights of hand, such as offsetting the center of some orbits from the center of the Earth.

Ptolemy provided precise predictions of planetary motion in his greatest work, commonly known as the *Almagest*. (Its real title was the *Syntaxis; Almagest*, which roughly translated means "Greatest Book," is the Arabic title of the text translated from Greek around the year AD 800.) The *Almagest* described the heavens with extraordinary mathematical sophistication. Its whole purpose, in fact, was to account for the motions of the lights in the sky—to save the phenomena. Ptolemy's system told you where the heavenly bodies would be and when, without worrying about what was really going on up there. Ptolemy described the physical aspects of his system in another book, *Hypotheses of the Planets*.

LET THERE BE LIGHT

In Grosseteste's day, Ptolemy's *Hypotheses of the Planets* had not yet been translated into Latin, although aspects of it were known via other writers. It was in any case Aristotle's model of the cosmos that most occupied philosophical inquiry. Grosseteste came of age just as Aristotle's works (mainly translated from Arabic) became widely known in Western Europe, and it was Aristotle's cosmos that Grosseteste studied, as presented in *De Caelo*, considered the authoritative guide to the physics of the heavens.[17]

Grosseteste approached cosmology not as an astronomer, interested in saving the phenomena, but as a natural philosopher. While philosophers such as Aristotle had inquired into physical explanations of heavenly motions, Grosseteste took a major step beyond. He wanted to figure out how all that cosmic machinery came into existence in the first place. That was no problem for Aristotle, who thought the cosmos was eternal and everlasting, without beginning or end. Grosseteste, though, operated in the theological framework of Christianity, in which God had created the cosmos— by letting there be light. Grosseteste's motivation was to reconcile the Aristotelian view with the scriptural importance of light that Augustine had emphasized.[18]

Grosseteste believed, correctly, that light emitted at a point emanates outward in all directions spherically. Just as a stone dropped into a pond propagates concentric waves away from the impact, striking a match sends light waves outward in an ever-expanding spherical shell. "Light of its very nature diffuses itself in every direction in such a way that a point of light will produce instantaneously a sphere of light of any size whatsoever, unless some opaque object stands in the way," Grosseteste wrote in his treatise *De Luce* (*On Light*), probably written around 1225.[19] That feature of light suggested to him an idea with at least a superficial similarity to the modern Big Bang theory of the universe's birth. Dimensionality—that is, three-dimensional space itself—might have come into existence by the action of light expanding from a point.

At the core of Grosseteste's theory for explaining existence in terms of light is the Aristotelian concept of form. Aristotle believed that pure matter (from which all material things are made) was formless, without meaningful existence. For Aristotle, bodily existence—or corporeity—requires matter plus form. Form imbues formless matter with properties that dictate its nature, and therefore its behavior. Matter possesses only the potential to become a real body; objects with sensible properties (that is, properties apparent to the senses) are always combinations of form and matter.

Grosseteste believed that pure matter did in fact possess a basic form conferring dimensionality or extension, making it a body—that is, giving it corporeity. This *forma corporeitatis* was conceived by the Persian Islamic philosopher Avicenna (980–1037), who is probably the source of Grosseteste's idea.[20] Grosseteste, though, was apparently the first to suggest that the source of corporeity—the prime corporeal entity—was light.

Both light and matter start out without three-dimensional existence, Grosseteste reasoned, but light is able to generate corporeity by multiplying itself infinitely as it spherically expands. In the beginning, then, light burst the universe into existence and drove matter outward—light's 3-D expansion somehow took matter along for the ride. Light, Grosseteste wrote, extended matter "along with itself into a mass the size of the material universe."[21]

As matter spread outward, its density would diminish. But it could not diminish forever, because that would lead to zero density—a vacuum, which nature (everybody then believed) abhorred. At some minimum density, matter's expansion could continue no longer. At that point matter took the shape of a spherical shell, with a "perfect" ratio of light to matter. Grosseteste called that first shell the firmament.[22] Inside the firmament, matter remained in a fairly ill-formed state, allowing "the possibility of further rarefaction."

Light from the firmament, Grosseteste proposed, shined inward, back toward the center of the universe. In the reverse of the original process, the inward-shining light dragged some of the leftover matter downward to create a second spherical shell, smaller and denser than the first. And then again, light from the second shell, shining inward, would drag matter with it to create another sphere. Over and over again, this procedure would create new spheres.

Eventually, though, light would fail to produce more spheres. Light from the innermost sphere, containing the moon, apparently did not possess enough corporeal power to create another perfect sphere. It did, though, produce light-giving fire, from which the other elements—air, water, and earth—then formed as well.

"This process . . . continued in this way until the nine heavenly spheres were completely actualized and there was gathered together below the ninth and lowest sphere the dense mass which constitutes the matter of the four elements," Grosseteste wrote.[23] All this leftover matter, creating the Earth as the central sphere, existed in an imperfect state—subject to change, while the nine heavenly spheres were immutable, impervious to corruption.

Grosseteste told a nice story, but from a modern view it lacks quantitative specificity. It wasn't that he didn't appreciate math; he was one of the earliest medieval advocates for the benefits of applying math to nature. "All causes of natural effects must be expressed by means of lines, angles and figures, for otherwise it is impossible to grasp their explanation," he wrote.[24] His belief in light as the source of corporeity led him to consider optics as the most fundamental science, and optics could be understood only through the geometric mathematics of lines and angles. "Grosseteste's conception of the nature of physical reality," Crombie wrote, "led him to the belief that nature could be properly understood only through mathematics."[25]

But Grosseteste's mathematical toolbox offered nothing capable of quantifying his elaborate account of cosmic creation. His description does, however, include enough detail that a modern-day scientist, equipped with a laptop and the right software, could translate Grosseteste's words into equations. In fact, a few years ago, an international team of medievalists, linguists, and other scientists took on just that challenge. They discovered that Grosseteste's theory really is amenable to mathematical rigor. "Grosseteste's model of how light interacts with matter . . . can, indeed, generate his claimed structure of the Universe," wrote physicist Richard Bower and colleagues.[26]

A GOLDILOCKS COSMOS

Obviously Grosseteste's universe does not resemble the version portrayed by modern cosmologists. But that's not the issue. Just as today's universe is described by the math of Einstein's general theory of relativity, Grosseteste's universe is also capable of intelligible mathematical expression. Bower and collaborators' math incorporates several subtle complications, but it's based on just a few key quantities. One—the central issue—is the strength of the interaction between light and matter. Other factors include the light's intensity and the matter's opacity. Bower's team identified those factors, composed equations to quantify them, and programmed a computer to solve those equations.

A first computer run seemed promising, except that it produced too many spheres. (Grosseteste was concerned only with the primary spheres, not the additional ones needed to guide the complicated planetary motions. So there should have been nine spheres beneath the firmament.) But that result was based on the assumption that the spheres were all completely transparent. Grosseteste's theory did not require transparency. If the interior spheres were less than perfectly transparent, the intensity of the light would diminish as it got closer to Earth and succeed in producing Grosseteste's 10 spheres.[27] "This is the type of universe in which Grosseteste imagines we live," Bower and colleagues wrote.[28]

There's a curious catch to this success, though. Equations implementing Grosseteste's theory produce the proper number of spheres only when the key quantities lie in a narrow range of values. Light intensity must be very high, for instance. The interaction strength between light and matter must be at just the right level. The opacity of the spheres must be high as well. It's a Goldilocks cosmos—everything must be just right to make it work.

It's in this regard that Grosseteste's cosmology resembles today's. To produce the cosmos that astronomers observe—with galaxies and stars, planets and people—certain physical quantities must be fine-tuned to values that aren't easily explained. Most prominent among these quantities is the strength of the dark energy, the mysterious force driving the cosmos to expand at an ever-accelerating rate. Had dark energy's strength been much greater, the universe would have expanded too rapidly, diffusing matter so thinly that stars and galaxies would have been unable to form. Weaker (or no) dark energy, combined with a sufficiently high density of matter in the cosmos, and the Big Bang would have fizzled—gravity would have collapsed the cosmos before there was time to make stars and galaxies at all.

As you'll remember from chapter 1 (you're only in chapter 2 after all), a possible explanation for the coincidence allowing the cosmos to exist as we know it is the multiverse. If an infinite (or very large) number of universes exist, each with different combinations of values for the key quantities, one will possess the right combination for life. In a similar way, Grosseteste's cosmos was the one where key quantities conspired to produce the right number of spheres. With different values for those quantities, his mechanism could have produced a whole range of universes with different numbers of spheres, just as some of today's cosmological theories imply the existence of a multitude of universes with differing features. Grosseteste did not realize this, of course. But even in hindsight, it's curious that a proposed mechanism for creating the universe could have created others, if certain properties (like the opacity of the spheres) could differ in value.

In Grosseteste's era, scholars did sometimes consider the possibility that the universe was not alone; they debated whether a "plurality of worlds" existed. Majority opinion, following Aristotle, argued that there could be only one universe. But Aristotle had also argued that the universe was eternal and everlasting, not created by some physical process, as Grosseteste had described in his effort to validate the creation story in Genesis.

Grosseteste did not address the plurality-of-worlds debate. But his theory is nevertheless important to the multiverse story—as an illustration. A creative thinker, eager to explain how the universe came to be, imagined a sophisticated physical process that could produce the known properties of the cosmos (at least as it was conceived at the time). And then a deeper analysis (much later) discovered that the same process could in principle have produced many other universes.

Just as Grosseteste's explanation for how the universe was built implied the possibility of a multiverse, so too does today's best theory of cosmic origins—inflation. If inflation burst our cosmic bubble into existence, it could have created other bubbles as well. Inflation makes sense of the universe we see, and a theory that explains what can be observed often predicts the existence of things that cannot be observed—in this case, other universes.

Grosseteste himself did not pursue the multiverse idea; he was predisposed to consider the existence of one universe only, as Aristotle had taught. Freed from that prejudice, his theory could have been adapted to accommodate the opposing ancient belief, advocated by the atomists, of a plurality of worlds.

Bower and colleagues could not say what Grosseteste thought about whether there might be multiple actual universes. Perhaps he might have considered the possibility in principle, even if only one actually existed. "We cannot know Grosseteste's view," wrote Bower and collaborators.[29]

Later in his life, though, Grosseteste did provide a clue to what he might have thought, in his theological writings while Bishop of Lincoln. Grosseteste believed that "nothing obliged God to make every feature of his creation evident to the human senses, with the result that there might well exist heavenly bodies that no one would ever see."[30]

That attitude at least suggests that Grosseteste's mind would have been open to the possibility of multiple worlds. Aristotle's was not.

Aristotle versus the Atomists

At the very birth of Astronomy, when the Earth was first asserted
to be Spherical, and to be surrounded with Air, even then there
were some men so bold as to affirm, that there were
an innumerable company of Worlds in the Stars.

—CHRISTIAAN HUYGENS, *COSMOTHEOROS,* 1698

SAINT JEROME did not paint a very flattering picture of
the poet Lucretius.

Born in northern Italy in AD 347, Jerome was a priest-historian
famous for translating the Bible into Latin (and he became the
patron saint of translators and librarians). In translating historical
tables (composed by his father Eusebius) from Greek into Latin,
Jerome occasionally inserted comments of his own. Some provided
biographical information about Roman writers. One such writer,
Titus Lucretius Carus, earned a curious entry in the account of the
years 96–93 BC.

"The poet Titus Lucretius was born," Jerome wrote. "Later he
was turned mad by a love potion, but in the intervals in between
the madness he composed some books, which Cicero afterwards
edited. He killed himself when he was 44 years old."[1] Among those
books, it turns out, was one of the grandest epic poems composed
in ancient Rome—*De Rerum Natura,* Latin for *On the Nature of
Things* (or "of the Universe").

Jerome lived four centuries after Lucretius, so couldn't have known for sure whether those stories about him were true or fake news (although Lucretius apparently did die young). It may well have been that describing Lucretius in an unfavorable light was a way to discredit his beliefs, which ran contrary to Christian dogma. For Lucretius denied that the world had a created beginning. And he attributed the workings of the world not to divine purpose, but rather to the mindless motions of an infinite number of invisibly small particles known as atoms.

Lucretius did not invent the atomic idea. He was a poet, not a physicist. But his account of the beliefs of the early Greek philosophers who proposed the existence of atoms offers by far the most complete exposition of ancient atomic theory that survived into modern times.

In Lucretius's day (and long after), the atomic theory of fundamental reality remained a minority view among scholars. Nowadays, of course, the reality of atoms has been established beyond doubt. But another aspect of the atomist philosophy espoused by Lucretius and his predecessors remains as controversial today as it was then: the existence of multiple universes.

ATOMS AND VOID

In the ancient world, rumors spread that the atom was conceived in Phoenicia before the Trojan War (sometime around 1200 BC). Credit was assigned to someone named Mōchus. Early modern atomists (such as Robert Boyle) believed that legend, but scholars since then have concluded that the Mōchus story was concocted long after the actual origin of atomism in ancient Greece. Today the prize for first atomist is commonly awarded to Leucippus, who lived in the fifth century BC.

As with many of the ancient Greek philosophers, details of Leucippus's life and work are fuzzy.[2] He may have been born in Miletus (or possibly Elea) but later taught in Abdera, a town in Thrace on the coast of the Aegean Sea, where he encountered the more famous early atomist, Democritus. Democritus was probably a student of Leucippus, and it's impossible to be sure which of Democritus's ideas were his rather than his teacher's. But ancient authorities such as Aristotle and Theophrastus believed that the basic outlines of atomic theory, at least, had been developed first by Leucippus.

Inspiration for the atomic idea perhaps came to Leucippus while he grappled with the philosophy of Parmenides of Elea (c. 515–450 BC). Parmenides taught that change and motion were impossible, reasoning that something that exists cannot emerge from something that doesn't exist. Therefore an object with a location, for instance, could not move to a new location, because that would bring a nonexistent location into existence. For some reason this was widely regarded as a powerful argument; its full force was illustrated by Parmenides's student, Zeno of Elea, in his famous paradoxes. Go ahead and try to shoot an arrow, Zeno challenged— it can't fly to the target, because first it would have to get halfway there, and before that, halfway to half way. Soon you realize that it can never move at all. Motion was therefore an illusion.

Modern scientists are baffled by this disregard for observational evidence. But among the early philosophers before Socrates (the pre-Socratics), reasoning of this nature was assumed to be a superior method of acquiring knowledge about reality. Still, Steven Weinberg, a Nobel laureate physicist who has written extensively on the history of science, nevertheless wonders why this view was so influential.

"What is most striking is not so much that Parmenides and Zeno were wrong as that they did not bother to explain why, if motion is impossible, things appear to move," Weinberg wrote.[3]

Perhaps Leucippus (who possibly was a student of Zeno's) was the Weinberg of his day. For Leucippus too found it crazy to deny the obvious reality of motion. He attacked a key defense for the Parmenides-Zeno philosophy: the impossibility of a void. Those who denied the possibility of motion argued that motion required the existence of a void. But a void is nothing, and nothing by definition doesn't exist. Therefore, there's no such thing as motion. Leucippus thought that through a little further and turned the reasoning around, concluding that since motion obviously existed, void must exist, too. And so he developed a theory to explain motion and change—by way of atoms moving in the void—in reply to the Parmenidean-inspired paradoxes posed by Zeno.

At least that is the traditional view, based on the writings of Aristotle. As Aristotle commented, quoting (probably paraphrasing) Leucippus: "That which is strictly speaking real is an absolute plenum; but the plenum is not one. On the contrary, there are an infinite number of them, and they are invisible owing to the smallness of their bulk. They move in the void (for there is a void); and by their coming together they effect coming into being; by their separation, passing away."[4]

Following Leucippus, Democritus also contended that what is real (plenum) exists, and so does what's not real—the void. As long as the plenum is not "one," but rather consists of innumerable tiny pieces, those pieces—the atoms—can move about and rearrange themselves to produce the changes in the world that everybody can plainly see. There was no need for nothing to become something. You simply needed to assume an infinite number of particles were surrounded by emptiness, a void, so the particles could move and assemble in various configurations conforming to sensible objects.

Democritus expressed this thought concisely in one of the passages from his writings often quoted by later scholars: "By convention sweet, by convention bitter, by convention hot, by convention cold, by convention color. But in reality atoms and the void."[5] That is, the objects of sense are defined by convention, taken as real but in truth merely manifestations of the atoms and void.

Leucippus and Democritus's atom isn't much like the one dissected by twentieth-century physicists, though. Today's atoms are swirling tempests of electric charges and energetic quarks, described by complicated equations. A Greek atom was solid and indivisible—the source of its name (from the Greek *atomos,* meaning indivisible or uncuttable). It's not entirely clear whether the Greek atoms had no parts or had parts that just could not be torn asunder. But atoms did have shapes, which accounted for the different properties of various substances.

For Leucippus and Democritus, an atom's shape was like today's atomic number—it gave the atom its identity. Atoms "have all sorts of shapes and appearances and different sizes," wrote the commentator Simplicius of Cilicia (c. 490–c. 560), describing the Greeks' ideas. "Some are rough, some hook-shaped, some concave, some convex and some have other innumerable variations."[6] Spherical atoms generated smoothly flowing substances, like water; scratchy stuff was composed of atoms with barbed extensions. Atoms could also differ in size—some scholars think Democritus would have allowed some atoms to be huge, but for the most part they were believed to be far too small to be seen.

Such atoms could make up everything that existed by virtue of their ability to move and connect to each other. Different combinations of shape, size, and orientation could form countless different sorts of objects, just like the letters of the alphabet can make all kinds of different words. Aristotle used that analogy in describing the atomic theory, pointing out that letters like A and N differ in shape and could make the two compounds AN or NA (different order), but could also make the two compounds ZA and AZ, because Z is the same as N, just in a different orientation.

INNUMERABLE WORLDS

Most cursory accounts of Greek atomic theory don't venture into cosmology. But just as today, when the physics of elementary particles is a core part of the science of the early universe, the Greeks saw a connection between atoms and the cosmos. Leucippus apparently realized from the beginning that making his system work required an infinite number of atoms. (Random agglomeration could not create a complex, structured cosmos from a limited number of atoms—you needed an inexhaustible supply, of which only a portion would create cosmic structure.[7]) And an infinite number of atoms implied the need for void that was also infinite. Leucippus composed a "Great World System" describing the creation of the world from the infinity of atoms traveling through that infinite void.

Diogenes Laertius, who during the third century AD composed biographies of the ancient philosophers, described Leucippus's explanation for how the world—Earth, planets, and stars—came into being. First, bodies of various shapes "by abscission from the infinite" move into the void, there producing "a single whirl, in which, colliding with one another and revolving in all manner of ways, they begin to separate apart, like to like." The "fine" particles escape from the whirl outward into the void, while others aggregate into a spherical structure. As the bodies in this sphere, or "membrane," continue to whirl, atoms nearest to the middle form the Earth. The membrane's outer edge attracts more atoms to form "moist and muddy" structures that dry out as the whirl revolves "to form the substance of the heavenly bodies."[8]

Leucippus did not believe that such a process happened only once. Theophrastus, Aristotle's successor at the Lyceum in Athens, ascribed to Leucippus the belief that from the atoms there arose "innumerable worlds." From an infinite number of atoms, an infinity of worlds would be created ("worlds" meaning the same as universes, or in Greek, *kosmoi*). As Diogenes wrote, "Just as there are comings into being of the world, so there are growths and decays and passings away in virtue of a certain necessity."[9]

Democritus agreed with Leucippus about the formation of the world and that there are many of them. While hints about the possibility of other worlds can be found in the ideas of earlier philosophers, none were as explicit and cogently formulated as those of the original atomists.[10] "Leucippus and Democritus," according to scholars Geoffrey Kirk, John Raven, and Malcolm Schofield, "are the first to whom we can with absolute certainty attribute the concept of innumerable worlds."[11]

It was central to Democritus's view that the innumerable worlds created by these atomic whirlpools would not all be alike (since the process creating them was random). As Diogenes related, Democritus believed those worlds would differ in size and other aspects. "In some worlds there is no sun and moon, in others they are larger than in our world, and in others more numerous. The intervals between the worlds are unequal; in some parts there are more worlds, in others fewer; some are increasing, some at their height, some decreasing; in some parts they are arising, in others falling. They are destroyed by collision one with another. There are some worlds devoid of living creatures or plants or any moisture."[12]

If some worlds were devoid of life or moisture, conditions necessary for life apparently did not automatically arise when a world was created. Modern views of the multiverse echo this reasoning. If the process of inflation produces multiple bubbles, some (at least one, anyway) would be hospitable to life while others would be lifeless.

BE HAPPY

While the idea of atoms was well-known in ancient Greece, it did not prevail as the dominant philosophical view of reality, thanks largely to Aristotle. Aristotle opposed atoms. While he clearly respected the depth of thought that went into the idea, it just didn't mesh with his own view of the world. Leucippus and Democritus, Aristotle pointed out, really had no good explanation for why atoms moved. And they did not specify what kind of motion it was. Was it the circular natural movement of heavenly matter? The linear natural motion of earthly matter? Or was this atomic motion unnatural, impelled to deviate from linearity or circularity by some constraint? "They say there is always movement," Aristotle complained, "but why and what this movement is they do not say."[13]

And of course Aristotle did not believe in the void, which was necessary for atoms to be able to move to begin with.

But atomism did have its champions. Most noteworthy and influential of them was the moral philosopher Epicurus.

Much of what is known about the life of Epicurus comes from Diogenes Laertius's biographical account.[14] Diogenes incorporated some of the letters Epicurus had written, including one to an acquaintance named Herodotus (not the famous historian) in which the Epicurean version of atomic theory is outlined. (A more complete work, *On Nature,* did not survive.)

Epicurus was born on Samos in 341 BC, son of an Athenian schoolmaster (Samos was an Athenian colony). When he turned 18, Epicurus traveled to Athens for compulsory military service. After that he moved around a lot, setting up schools of his own on the island of Lesbos, later at Lampsacus (a Greek settlement on the Asian side of the Hellespont), and finally back in Athens. His school there, called the Garden because it met in the garden at his house, espoused a new philosophy of life and nature very much unlike those at competing Athenian schools.

Epicurus was an advocate of happiness. He would have appreciated the line "happiness is the truth" from the 2013 Pharrell Williams song "Happy," or perhaps would have preferred Bobby McFerrin's 1988 number-one hit "Don't Worry, Be Happy" (which was actually the slogan of the twentieth-century Indian mystic Meher Baba).[15]

Along with the hedonistic admonitions to seek pleasure and avoid pain, Epicurean philosophy advised isolation from the worries of the world and recommended "living in obscurity." That was the road to peace of mind. But you could not live a hedonistic life and retain your peace of mind while fearing punishment in the afterlife from moralistic gods. And so Epicureanism required the comfort of a philosophy of nature in which gods played no part. Conveniently, just such a cosmological picture had been provided by the atoms of Leucippus and Democritus.

Epicurus did not deny the existence of gods; he merely asserted that they didn't bother with human affairs. Reality was ruled by the properties of atoms. They might be imperceptible to the senses, but nevertheless their existence could be affirmed by reasoning. "The universe is bodies and space," Epicurus declared. Without void and place, "bodies would have nowhere to exist and nothing through which to move, as they are seen to move." Among bodies, some are compounds, which implies the existence of smaller bodies from which compounds can be formed. Those bodies must be "indivisible and unalterable." Otherwise nothing would remain when compounds dissolved, and all existing things would eventually succumb to destruction into nonexistence. Those permanent unalterable bodies are "completely solid in nature, and cannot be dissolved in any part." They are "indivisible corporeal existences"—for short, atoms.[16]

Regularities in the movement of heavenly bodies, such as risings, settings, and eclipses, are not the work of any being who controls or ordains them. Regularity in cosmic motions is rather a result of how atoms moved and came together to make the world in the first place. "We must believe that it is due to the original inclusion of matter in such agglomerations during the birth-process of the world that this law of regular succession is also brought about," Epicurus proclaimed.[17]

For the most part, Epicurus adopted the views of Democritus on the nature of atoms. But Epicurus added at least one important tweak, known as "the swerve." As described by Democritus, atoms simply moved about randomly—the motion of one was determined by its impacts with others. Intrinsic atomic motion was just accepted as a basic principle of the theory. As Simplicius described the original atomists' view, "these atoms . . . move in the void, overtake each other and collide. Some of them rebound in random directions, while others interlock because of the symmetry of their shapes, sizes, positions and arrangements, and remain together."[18]

Epicurus, on the other hand, believed atoms moved because they had weight and therefore constantly "fell downward" through the void. But either way—intrinsic motion or downward motion—it was hard to find a place for human choice. Either human-induced motion was just an artifact of the random collisions of atoms, or atoms wouldn't even create humans but rather would all just fall straight down to the center of the cosmos. And so Epicurus introduced the swerve: an unpredictable alteration in atomic motion that would eliminate the deterministic chain of collisions. The swerve opened a door for complicated structure to form and for human free will to influence events.

It's not easy to imagine how this swerve actually worked. But it curiously calls to mind some modern attempts to use the unpredictability inherent in quantum mechanics to allow human free will into the universe.[19] It would be going much too far, though, to suggest that Epicurus had any inkling of quantum physics.

While modifying the atomic theory of Democritus in some respects, Epicurus retained the cosmological implication of multiple universes. "There are infinite worlds both like and unlike this world of ours," he wrote to Herodotus. "For the atoms being infinite in number, as was proved already, are borne on far out into space." Among these "worlds created from the infinite," Epicurus believed, some would be expected to include "living creatures and plants and other things we see in this world," as no one could prove that the seeds for such things would not have been formed there.[20]

Epicurus died in 270 BC. As with Democritus and Leucippus, the nuances and depth of Epicurus's thought as expressed in his own words are lost. But in the case of Epicurus, many details did survive, in the epic poem of his devoted follower Lucretius.

WHERE SIGHT CAN NEVER SOAR

Apart from Saint Jerome's cryptic account of Lucretius's life, very little is known about him.[21] If he was born around 94 BC, as Jerome's chronology suggests, then Lucretius died about 50 BC or thereabouts (if Jerome is right about the age of death). In 54 BC the Roman politician and orator Cicero (mentioned by Jerome as Lucretius's editor) wrote a letter commenting on the "flashes of genius" that appeared in Lucretius's poems.[22]

Whether Cicero actually edited *De Rerum Natura,* though (perhaps polishing it up after Lucretius's death), is not so clear. And Jerome's depiction of Lucretius as psychotic and suicidal has been widely criticized on the grounds that it reflected Christian disapproval of Epicurean philosophy, demanding that Lucretius be portrayed as unadmirable.[23]

In any case, Lucretius's famous poem was no doubt written before 50 BC, and it made quite an impression. Virgil (70–19 BC), the more famous Roman poet, incorporated glowing praise of Lucretius in one of his works. Other poets imitated Lucretius's style. Even though he practiced the Epicurean preaching to live in obscurity, Lucretius's poetry was well-known in ancient Rome. And it remains to this day an exemplar of techniques for communicating complicated philosophy (even science) to a general audience.

Consider his example of the statue. While in Roman times nobody used the phrase "high five," it was customary when walking by a statue to reach out and grasp the statue's right hand. As years passed by, the wear on the right hand (compared with the left) became apparent. Yet of course, no one handshake produced any visible effect. Lucretius argued eloquently that something must be removed from the hand with each touch. That which was removed must have been too small to see, a truth easily explained if everything was made from invisibly small atoms.

For Lucretius, nature is full of illustrations supporting (or even demanding) the existence of atoms. Statue hands provide just one of many such examples. Similarly a ring on a finger grows thinner as the wearer ages. Paving stones on roads are rubbed down by foot traffic. And everyone can see the power of the wind to propel a sailboat or blow down a building. What else other than invisible particles could accomplish such feats? When water causes destruction similar to the wind's, we can see the liquid in action. But what about clothes hanging near the sea—they become damp, while initially moist garments hung outside to absorb sunlight dry out. You can't see the water arriving or leaving. "The moisture, therefore, is dispersed into minute particles, which our eyes can by no means perceive."[24]

In short, evidence for the existence of atoms is all around us. And by observing the world closely, we can discern some of their properties.

It was important for Lucretius (as for Epicurus) to demonstrate that all of the perceived world came into being from the actions of atoms. Lucretius argued that the atoms certainly did not conspire by their own free will, but rather buffeted themselves about—"for an infinite space of time agitated"—into all manner of possible configurations, eventually forming the arrangement corresponding to the stars, sun, and Earth with its "breathing creatures and blooms"—the "sum of things." This arrangement of atoms and their motions was such that nature's behaviors—the flowing of rivers, the sun's heat, the flourishing of living creatures—were sustained by the arrival of new atoms from infinite space. And if this version of reality be true—as Lucretius believed he had demonstrated—then the existence of other worlds, other universes, was established as well. The world could not work as it did unless "an abundant supply of matter could arise from the infinite of space."[25] For just as animals deprived of food waste away, the world's structures could not persist without a never-ending fresh supply of atoms.

Having made this point, Lucretius proceeds to tell Memmius (the Roman aristocrat to whom the entire poem is addressed) that the reality of an infinite space requires contemplation of what lies beyond the known cosmos. "My mind proceeds to make inquiry what there exists farther onwards, into those parts which the mind perpetually longs to look, and into which the free effort of thought itself earnestly desires to proceed." Or as a more poetic translation puts it: "Since space spreads boundless, the redundant mind, free in its flights, pants, ardent, to discern what fills those realms where sight can never soar."[26]

And where sight can never soar, Lucretius believes, other worlds must exist. When atoms "of incomputable number and unfathomable sum . . . fly through the void in infinite ways," he declares, "by no means can it be thought probable . . . that this one globe of the earth, and this one heaven, have been alone produced." With an abundance of matter, and an infinity of space, natural causes—atoms acting on their own—will repeatedly produce the complexities of nature, Lucretius avers. "If there is such a vast multitude of seminal-atoms as the whole age of all living creatures would not suffice to number, and if there remains the same force and nature, that can throw together the atoms of things into every part in the same manner as they have been thrown together into this, you must necessarily suppose that there are other orbs of earth in other regions of space, and various races of men and generations of beasts."[27]

As these passages indicate, a multiplicity of worlds was not merely an embellishment tacked on to the atomists' primary proposal that matter is composed of tiny parts invisible and indivisible. Only an infinite number of atoms could have created the complexity of the known world by their random motions. And no matter how many atoms composed the known world, an infinite number remained to configure themselves into other worlds. So if matter is made of atoms, then an infinite number of other worlds must exist—the atomists' multiverse is a consequence of their atomic theory, not a separate proposition. "The belief in a vast number of worlds," wrote physicist-historian Samuel Sambursky, was an "integral part of the atomic cosmogony."[28]

This point had already been grasped by Leucippus. An infinite number of worlds was also for him a consequence of the existence of atoms and void. Universes unseen must of necessity exist in order to explain the things observed in the known world. It didn't matter that those other universes were inaccessibly far away.

In this sense, the atomist-multiverse theory of antiquity presents a striking parallel to the situation in science today. The Greek atomists' theory of the ultimate nature of matter on the smallest scales implied the existence of multiple universes on cosmic scales. Modern science's most popular attempt to describe the fundamental nature of matter—superstring theory—also turns out (much to the theorists' surprise) to imply a vast multiplicity of vacuum states, essentially the same thing as predicting the existence of a multiverse. That parallel applies also to the negative reaction of skeptical scientists: Aristotle's rejection of the atomists is reflected today in widespread opposition to (or at least disregard for) superstring theory. And just as Aristotle rejected a multiverse, so too do many scientists today. Much as with ancient (and medieval) scientists or philosophers, whether there is one universe or many is today still a topic of vigorous debate. Now as then two very different views of ultimate reality compete for scientific respect. A question raised millennia ago remains today a central point of contention among modern scientists attempting to comprehend existence.

For many centuries between the ancient and modern eras, though, the insights of Leucippus and Democritus, Epicurus and Lucretius remained irrelevant to whatever passed for the practice of cosmological science. Aristotle's contrary views won the hearts and minds of most ancient philosophers. Both Aristotle and his teacher Plato had concurred that there could be only one natural world—a nested set of spheres with the Earth at the center. Earth occupied the center, Aristotle insisted, because that was earth's (the element's) natural place. No other such world could exist, he argued, because more than one world would mean more than one center, so the element earth would not know where to go. It could not move both toward the center of another world and ours; that would pose a logical contradiction, and to Aristotle—the founder of formal logic—a logical contradiction was a personal affront. (Plato had affirmed the uniqueness of the cosmos on different grounds. It had to do with his concept of perfection, which he regarded as requiring "unity.")

It was this view of a single cosmos from ancient thinkers that passed to western European scholars, via Arabic-speaking commentators on Aristotle, during the epoch of reawakened intellectual activity in the later Middle Ages. Lucretius's poem, with its account of the Greek atomistic philosophy, was largely unknown during that rise of European scientific activity. Discussions of the possible plurality worlds among medieval philosophers were all framed in terms of commentaries on Aristotle's opposition to the idea.[29] Nobody asked who was right, Aristotle or Lucretius. It was always a question of whether Aristotle was right or wrong. Almost all the prominent thinkers ultimately sided with Aristotle. At least until the Bishop of Paris, acting with the backing of the pope, ordered them not to.

The Condemnation of 1277

The attempt to build a new physics at the university of Paris
in the fourteenth century laid the foundations of modern
science; it was born, one can say, March 7, 1277,
of the decree by Stephen, bishop of Paris.

—PIERRE DUHEM, *LE SYSTÈME DU MONDE*

THROUGHOUT THE LITERATURE on scientific history run diverse opinions about when modern science should celebrate its birthday.

Science of some sort, everybody acknowledges, was practiced in ancient Greece by the likes of such pre-Socratic philosophers as Thales of Miletus and his student Anaximander. But that was not *modern* science. It was philosophy infused with a curious concern for understanding nature on its own terms rather than as a reflection of the whims of various gods. Modern science, various commenters contend, began with Copernicus. Or Francis Bacon. Or Galileo. Some would hold out for Newton.[1]

But for the French philosopher-historian-physicist Pierre Duhem, modern science—or at least modern physics, which counts as basically the same—was much older than that. In the early twentieth century, Duhem undertook a thorough investigation of the science, such as it was, that was practiced throughout the Middle

Ages. His verdict: modern science received the spark of life in 1277, when the bishop of Paris ordered the scholars of the university there to cease teaching various tenets of Aristotle, under penalty of excommunication. "One can say that the excommunications pronounced at Paris on March 7, 1277, by Bishop Étienne [Stephen] Tempier and the doctors of theology were the birth certificate of modern physics," Duhem asserted.[2]

More recent historians insist that Duhem's pronouncement was severely exaggerated. But as David Lindberg noted, it was "not entirely wrong."[3] In any case, there is no doubt that Tempier's edict ignited fresh discussion of an issue that has fascinated (and divided) scientists ever since: the possibility of a plurality of worlds.[4] That was one of the propositions that Aristotle had denied.

Even long before European scholars had become enamored with Aristotle, they mostly maintained the belief in a single cosmos. Throughout the earlier Middle Ages, conventional cosmological wisdom was based on Plato's conception of the cosmos as described in his *Timaeus*. And Plato had insisted on only one universe. In fact, he referred to it as a single living being. "For the god, wishing to make this world most nearly like that intelligible thing which is best and in every way complete, fashioned it as a single visible living creature, containing within itself all living things," Plato contended.[5]

Plato acknowledged that others had suggested the existence of an indefinite number of worlds. But he found that view inconsistent with his own concept of the "body of the world" as all-inclusive, whole, and complete. "For that which embraces all the intelligible living creatures that there are, cannot be one of a pair; for then there would have to be yet another living creature embracing these two, and they would be parts of it," he wrote. "For that reason its maker did not make two worlds nor yet an indefinite number; but this Heaven has come to be and is and shall be hereafter one and unique."[6]

Then in the twelfth century, Aristotle invaded western Europe in the form of translations from Arabic, inspiring a renewal of intellectual activity in various European locales. By the early thirteenth century, novel institutions—universities—had begun to sprout across the continent. Under the new systematic approach to higher education, investigation of the universe took the form of debating questions raised by Aristotle's works and composing commentaries on them. Those debates and commentaries incorporated influences from the Aristotelian writings by Muslim philosophers, most notably Ibn Rushd (1126–1198), better known in Christian Europe by his Latin name, Averroës. As Aristotle was commonly referred to as The Philosopher, Averroës was designated The Commentator.

In Aristotle's treatise *De Caelo*, he argued strongly—in opposition to the atomists—for one world alone, one cosmos, a set of concentric spheres outside of which nothing existed. And by nothing he meant nothing, not even a vacuum or void. To Aristotle, asking what was outside the outer sphere was like a scientist today asking what on Earth is north of the North Pole. It's a question without meaning. It's like asking who won the baseball World Series in 1994; there was no World Series in 1994 (players on strike). And there is no meaning to asking what's outside the universe. The universe is everything.

Robert Grosseteste, as you'll recall from chapter 2, based his own cosmological investigations on *De Caelo*. (Late in his life, Grosseteste even undertook a translation of *De Caelo* from the original Greek, in order to avoid any distortions possibly introduced by the Latin translations from Arabic.) Grosseteste did affirm Aristotle's contention that a vacuum could not exist. But he did not investigate Aristotle's rejection of multiple universes. Grosseteste's disciple Roger Bacon, though, was among several thirteenth-century scholars who addressed Aristotle's views on the plurality of worlds in detail.

DOCTOR MIRABILIS

Bacon's birthdate is uncertain; it was possibly as early as 1214 or as late as 1220. Old sources suggest that Bacon was a student of Grosseteste's, but modern scholars don't think so.[7] Bacon apparently did study at Oxford, though, and also at Paris; by the 1240s he was lecturing in Paris on some of Aristotle's works. About 1247 he returned to Oxford (long after Grosseteste had become Bishop of Lincoln) but was in Paris again in 1251. In 1257 he became a Franciscan friar, which didn't turn out to be an especially good fit. (Bacon seemed to be always getting in trouble for refusing to adhere to all the Franciscans' rules for beliefs and behaviors.)

Bacon's scholastic nickname, Doctor Mirabilis, attested to his creative intellectual power. He was famous among early medieval philosophers for an apparent ability to foresee the future. (His writings can even be interpreted as predicting such technologies as submarines and airplanes.) He also was lauded by later commentators because he was an early advocate for doing experiments and especially promoted the value of mathematics for understanding nature. Some researchers also credit him (citing his further development of Grosseteste's ideas) with introducing the idea of "laws of nature."[8] Aristotle had discussed phenomena in terms of substances acting according to their individual "natures"; Bacon argued for more general laws of nature that governed everything.

"Grosseteste and Bacon held that nature was ultimately mathematical and could be explained by mathematical laws," wrote Yael Kedar and Giora Hon. "They thus introduced, instead of the Aristotelian 'form' with its irreducible qualitative differences among different substances and their movements, a new concept of law of nature as the proper object of scientific inquiry."[9]

Whether he studied directly with Grosseteste or not, Bacon was certainly a big fan. Notoriously critical of many other famous names in the medieval intellectual world, Bacon profusely praised

Grosseteste. As one Grosseteste biographer noted, Bacon "was not inclined, as a general rule, to praise either his immediate predecessors or his contemporaries." On Albert the Great, for instance, Bacon affirmed that there was some in his writings that was useful, but it could have been summed up in a treatise "20 times as short" as Albert's actual works. And Bacon declared that Albert and Thomas Aquinas "both became teachers before they had been adequately taught" (a double slam on Albert, who was Thomas's teacher). On the other hand, Bacon remarked that "one man alone had really known the sciences, namely, Robert, Bishop of Lincoln. . . . The Lord Robert alone, on account of his long life and the wonderful methods he employed, excelled all men in his knowledge of the sciences."[10]

Like most medieval scholars, Bacon pursued issues of natural philosophy by examining questions drawn from Aristotle's writings, articulating the arguments pro and con, and then drawing a conclusion. Among the many questions commonly addressed by thirteenth-century scholars was the possibility of the existence of the void. While he denied that possibility, Aristotle had not linked the issue of the void to the prospect of a multiplicity of universes. But Bacon and other medieval scholars seized "no void" as a crucial point in the plurality-of-worlds debate. (No void, no place for those other worlds to be.) And Bacon was rather vigorously opposed to the idea of a void. Like Aristotle, Bacon insisted that nothingness—the void—was literally nothing, which meant it didn't, or couldn't, exist. And so when it came to the question of multiple worlds, Bacon's position on the void had in effect already provided the answer.

"If another world existed, it would have the shape of a sphere like this one," Bacon reasoned. Those two worlds could not be separate from one another, because that would imply a void, so they must touch each other. "But by the 12th proposition of the third book of Euclid's *Elements,* they could not touch at other than one point. . . . So everywhere other than that point, there would be a void space between them." And since voids are not permitted, there cannot be other worlds. "If there were several worlds, in fact, it would permit a void, which has previously been refuted in a general way."[11]

In this view Bacon echoed Scottish scholar Michael Scot (1175–1232), who also based his one-universe belief on nature's abhorrence of the vacuum. He relied on the geometric argument: an external sphere would require a vacuum separating it from our sphere, therefore no such external sphere could exist. "Therefore," Scot concluded, "there cannot be many worlds."[12]

William of Auvergne (d. 1249), who lectured in Paris during Bacon's stay there, expressed similar views. In a major work on the natural world that included a segment called *De Universo,* William insisted that there was, in fact, only one universe, also emphasizing the impossibility of a void. He did raise the interesting possibility of "another world" existing in a gigantic sphere beyond the known spheres of the cosmos, containing them within. But even if this were so, William argued, the heavens of that world and of our world would still all be just part of only one universe. "It is clear that this sphere, with all that it envelops, forms a unique world, containing in it all things," William wrote.[13] A century later Nicole Oresme would explore that idea much more thoroughly.

Bacon, Scot, and William of Auvergne all approached the plurality-of-worlds question strictly on Aristotle's terms. They took Aristotle's view as the starting point, rather than grappling with the actual arguments of the atomists. Addressing the question of multiple worlds in the Middle Ages was simply part of the process of producing commentaries on Aristotle (which was, in fact, basically the job description for medieval scholars). Bacon and the others were essentially trapped in Aristotle's paradigm; for them, the sphericity of the world was a given, from which simple geometrical arguments showed that another spherical world, capable of touching at only one point, required a void by virtue of spherical convexity. While medieval writers did occasionally question

Aristotle on some specific points, nobody seemed to doubt the big picture of concentric spheres forming the cosmos. All cosmological arguments were posed within that framework.

Modern scientists have often congratulated themselves on escaping from Aristotle's incorrect paradigm and replacing it with the modern "correct" view of the cosmos. But it seems to me that there are elements of modern understanding that also are simply taken for granted. Just as medieval scholars accepted Aristotle's nested spheres without question, many scientists today seldom examine their hidden assumptions. And much of the modern multiverse argument challenges those assumptions. In the twentieth century, for example, rarely did anyone question the obvious fact that space consisted of three dimensions. But the development of superstring theory raised the serious prospect that additional dimensions might have escaped anyone's attention. String theory's extra dimensions open a whole new set of possibilities for explaining natural phenomena (with one of the implications being the possible existence of a plurality of worlds inhabiting those other dimensions).

Bacon was no string theorist. His arguments against multiple worlds were firmly rooted in Aristotelian ground. Besides agreeing that there could be no void, Bacon approved Aristotle's view that the world contains all the matter there is. That makes the world unique and singular: there can be just one. Bacon also noted that if a second world could exist, then so could a third, a fourth, and on to infinity. "It must therefore be that there is an infinity of worlds or not more than one; but we know there cannot be an infinite number of worlds. Therefore, there is only one."[14]

Bacon's views reflected the early thirteenth-century consensus that Aristotle was right: there was only one world. That position was maintained throughout the middle decades of the century, articulated and affirmed by two of that era's philosophical giants, Albert the Great (Albertus Magnus) and Thomas Aquinas.

A COMPLETE EXPOSITION OF NATURAL SCIENCE

Albert the Great (as he was to be called even in his lifetime) overwhelmed his contemporaries with his prodigious scholarly output. Most of his fellow philosophers were very impressed, except of course for Bacon. As with many of the most prominent of medieval scholars, Albert earned a complimentary (or at least descriptive) nickname—in his case, Doctor Universalis. Albert's fame persisted throughout the centuries that followed his death, and in 1931 he achieved sainthood. A decade after that, Pope Pius XII proclaimed him the patron saint of researchers in the natural sciences.

Albert was born in Bavaria around the year 1200.[15] As a young man, he went to Italy and studied Aristotle (possibly at Padua). In 1223 he joined the order of the Dominicans and studied theology in Cologne. By 1240 the Dominican leadership sent him to Paris, the European capital of theological education, to earn his doctorate. Later in the 1240s, he taught at Paris, with Thomas Aquinas one of his students. Albert soon returned to Cologne (with Thomas) to teach and spent the rest of his years taking on various tasks in service of his order and the pope, including a short stint (1260–1262) as bishop of Regensburg.

Like Grosseteste earlier in his century, Albert was devoted to the reconciliation of theology and science. He applied the principles of reasoning articulated by Aristotle to both realms. In fact, Albert played a major role in introducing Aristotle to the rest of the Western scholarly world. "Albert deserves our attention and respect," wrote Lindberg, "as the man who put Western Christendom fully in touch with the Aristotelian tradition."[16]

Much of Albert's prodigious output consisted of commentaries on Aristotle's complete works (at least, as much of them as was available in Latin back then). Where Aristotle left scientific topics unmentioned, Albert wrote his own treatises to fill the gaps. Albert intended for his work to encompass the totality of reality. "Our purpose," he wrote in one of his commentaries, was to provide the members of his order "a complete exposition of natural science" enabling them to properly understand the works of Aristotle.[17]

Albert was a voracious reader of the philosophers who had come before him, from the other Greeks such as Plato to the Muslim philosophers including Avicenna and Averroës (The Commentator). Averroës was Albert's guide in formulating his assessment of the plurality-of-worlds question (which, Albert commented, was "one of the most wondrous and noble questions in nature").[18] On other matters Albert occasionally would disagree with Aristotle, but in this case Albert concurred with the one-universe conclusion. After articulating reasons for possibly rejecting Aristotle's arguments, Albert in each case poked various holes in the logic underlying the objections. Would earth (and other elements) actually be pulled in an opposite direction by the presence of another world, as Aristotle insisted? Some scholars argued no, suggesting that an element would still seek its natural place in the closer world. Our earth would still congregate beneath our feet because the center of the other world would be so much farther away. Defenders of this argument cited magnetism—a magnet could attract a nearby piece of iron to it but leave undisturbed any magnetic metals at great distance.

Such an argument, Albert maintained, did "not conform to the rules of reason, and consequently, it is erroneous."[19] Elements do not move by some force of attraction, but rather by an "intrinsic motive power." An element's form—that which is imbued when the element is created—defines that element's natural place and therefore its natural movement. If some attractive object infused some other form into the element, causing it to be pulled in a direction other than its natural one, that element would become a composite body, no longer an element.

"The coexistence of two such forms is impossible," Albert insisted. Therefore form must stay the same regardless of the object's distance from its natural place. "Either near or far from its natural place, it always moves with simple motion."[20] So basically, Albert simply bought Aristotle's argument, as presented by Averroës.

Albert's most famous student, Thomas Aquinas, took up the issue later in more detail. Thomas, born in Italy, learned about Aristotle while a student in Naples, became a Dominican friar, and eventually went to Paris to study theology, where he was taught by Albert.[21] Like Albert, Thomas argued for the mutual compatibility of faith and natural science (or natural philosophy, the preferred label in those days). Revelation from scripture and the natural world both emanated from God, Thomas pointed out, and therefore there could be no conflict between them. Both Albert and Thomas held, in common with other scholars of the day, that science was theology's handmaiden, a tool for discovering and understanding the glory of God's creation.

God's creation in its totality—the entire universe itself—was therefore a proper topic of scientific study (or natural philosophizing). And the question of whether the one known universe was in fact the totality of God's creation naturally elicited in-depth philosophical investigation.

Thomas viewed that question through the writings of the Aristotelian commentator Simplicius, agreeing with him (and Albert) that another universe's great distance would not help an element know where to go. "To postulate a difference of nature in the simple bodies according as they are more or less distant from their proper places is unreasonable," Aquinas wrote. "For what difference can it make whether we say that a thing is this distance away or that? . . . The form is in fact the same."[22] It's true that an object moves faster when it gets closer to its natural home—falling bodies do accelerate. But that's just a change in speed—a difference in magnitude of the motion—not a change in the nature of the body or its natural direction of motion.

Thomas recognized, though, that Aristotle's reasoning about natural place was not sufficient. Aristotle had (obviously) not considered the possible powers of the omnipotent Christian God. To some theologians, it didn't seem unreasonable for God to create more than one cosmos. Thomas, as was the custom for medieval scholars, took those arguments seriously and articulated them fairly. If God's power is infinite, the argument went, producing a single world would not attain the limits of his powers, so obviously he could have created other worlds. But to this proposal Thomas offered a clever response. If God were to make other worlds, they would either be similar to this world or different. It would not make sense for God to make multiple similar worlds, though, because "if entirely alike, they would be in vain—and that conflicts with His wisdom."[23] But if he made them different, then none would contain all that could be, and therefore none of them would be perfect. So on the issue of making multiple worlds, God's perfection conditions his omnipotence.

Another argument for the multiplicity of worlds invoked an analogy to the multiple individuals among types of living things. A noble genus (of horses or oxen, for example) comprises many different individuals—it possesses the power to be realized in many different ways. If a genus "is capable of making up a number of individuals, then *a fortiori* the genus of the universe can contain several individuals," as Thomas paraphrased the argument.[24] To this reasoning Thomas replied that individual horses or oxen are all imperfect—none contain "everything proper to its kind." It takes a greater power to create the single perfect individual—that is, a single perfect universe—than many imperfect ones. Once again, Thomas affirms the importance of omnipotent power to produce perfection.

But wouldn't the goodness of the universe be an even greater good if there were many of them? No, because the goodness of the universe is rooted in its unity. Division into a multiplicity would diminish the goodness of the unity of the cosmos, Thomas insisted. It was its unity that sealed the deal for the single cosmos: Thomas believed that whatever existed came from God. Therefore any two entities must be related both to God and to each other, a unifying force such that "it is necessary that all things should belong to one world." It was failure to acknowledge an "ordaining wisdom" that allowed atomists such as Democritus to infer "that this world, besides an infinite number of other worlds, was made from a coming together of atoms."[25]

Thomas became world famous and was long studied as the master of medieval theology and its merger with Aristotelian science. He died in 1274 (six years before his elder, Albert) and achieved sainthood only a few decades afterwards, in 1323. (Today Thomas is commonly listed as the patron saint of students and teachers.) But in his own time, some controversy surrounded Thomas's writings, and three years after his death many of his teachings were declared forbidden—including the declaration that there could be only one universe.

CONDEMNATION

Throughout the thirteenth century, tension often accompanied the interaction between the theologians and the natural philosophers (even though many individuals were both). Aristotle's science did not exactly correspond to the dogmas of Christian teaching. Despite the efforts of Bacon, Albert the Great, Thomas Aquinas, and others to reconcile Aristotle and Christian faith, mainstream theologians continued to assert that God's power would enable him to do things that Aristotle declared impossible.

Natural philosophy had begun to flourish in western Europe during the eleventh and twelfth centuries, as translations from Arabic of the Greek authors and their Islamic commentators flowed into a Latin-literate intellectual community "starved for knowledge."[26] That hunger for knowledge gave birth to a new type of institution, the university. By the start of the thirteenth century, prominent universities had been or were soon to be established in such leading centers of learning as Bologna, Paris, and Oxford. It was at the universities that natural philosophy established itself as a way of knowing reality—complementing, if not competing with, the knowledge of reality revealed by scripture.

Scholars reveled in the wealth of knowledge contained in Aristotle's writings. But church leaders and conservative theologians expressed concern that Aristotle's views didn't always mesh with Christian teachings. Aristotle, for instance, maintained that the world is eternal. He argued that a beginning or end to existence is illogical. Christian belief, of course, asserted a time of creation and forecast an ending. Aristotle also insisted that all the matter from which things took shape had existed beforehand, denying the possibility that something could be created from nothing. Which is just what the Christian God supposedly did when creating the universe.

Other details of Aristotle's philosophy annoyed certain theologians. In particular, he taught that some thinkable things were actually impossible, such as creating a vacuum or—as we've seen—creating another cosmos. To many churchmen, Aristotle seemed to be limiting God's power; they believed fervently that God could do anything that was not in and of itself logically contradictory. Richard of Middleton (c. 1249–c. 1302) made the point that the inability to make two contradictory things exist simultaneously was not a limit on God's power, but merely an observation reflecting the nature of the concepts of "affirmation and negation." It makes no sense to imagine a "power" that could change the meaning of those words. It's no limit on God to say he can't change the fact that a whole consists of parts, because "such is the nature of whole and part."[27]

Some natural philosophers attempted to cope with the theology-philosophy tension by arguing that some things could not happen "naturally," without denying that God could cause such things to happen supernaturally. That is, God could intervene to counter the natural order of things. Nevertheless the defenders of faith found Aristotle sufficiently unsavory that they instigated attempts to ban his works. In 1210 and 1215, religious authorities in Paris imposed such bans, but not very successfully. (For one thing, the ban applied only to the Paris arts faculty; theologians could still read Aristotle if they wanted to. And Aristotle was freely discussed at other universities, such as Oxford.) In 1231 Pope Gregory IX declared that Aristotle's works should not be banned, but rather merely "corrected." Since Aristotle's books contained "both useful and useless matter," the pope explained, it would be better just to weed out the content that was "erroneous or likely to give scandal or offense to readers." What remained could "be studied without delay and without offense."[28]

Supposedly a committee was established to identify and purge Aristotle's errors, but there is no sign that the committee ever got its act together enough to write a report. In any case, by the 1240s Bacon and others were delivering lectures on Aristotle at the University of Paris, and most natural philosophers across Europe had thoroughly assimilated Aristotle into their worldviews. In 1255 the Paris faculty documented a whole list of Aristotle's works that were required reading. It seemed that Aristotle was winning.

Still, the theologians, especially in Paris, disapproved of Aristotle on many points. But instead of banning him outright, they decided to focus more on condemning the specific propositions they found most distasteful. Thirteen such teachings were banned in 1270. Apparently that action was insufficient to deter the Aristotelians, though. So the author of the ban, Bishop of Paris Étienne Tempier, decided he needed to do the job more thoroughly.

During this time word of the heretical teaching by philosophers in Paris had come to the attention of Pope John XXI. In 1277 he instructed Tempier to investigate, but Tempier may have already set some anti-Aristotelian wheels into motion.[29] In any case Tempier convened a group of prominent theologians and consulted with "other prudent men" to produce a list of 219 articles—propositions relevant to the tensions between theology and philosophy—to be condemned. Anyone teaching any of those propositions (or even listening to them being taught) was subject to excommunication: "Lest dangerous discourse should draw the innocent into error, we strictly forbid . . . and wholly condemn these things, excommunicating all those who shall have taught some or all of the said errors, or shall have presumed to defend or support them in any way whatever, and also those who listen to these things," Tempier declared.[30] Anyone guilty of such acts had seven days to report to the bishop or the chancellor of Paris to plead for relief from excommunication, though they would still be subject to other possible punishments.

Most of the banned propositions reflected Aristotelian beliefs. Article 49 declared it an error to allege "that God could not move the heavens . . . with rectilinear motion," something Aristotle considered impossible. Article 87 prohibited teaching that "the world is eternal as to all the species contained in it," as Aristotle believed. Several other articles also condemned aspects of the world-is-eternal belief—Article 93, for instance, held it an error to teach that "celestial bodies have eternity of substance but not eternity of motion." But in addition to the anti-Aristotelianism articles there were a few condemnations of a more general nature, reflecting the theologians' concern with philosophers' audacity. It was now forbidden, according to Article 154, to maintain that "the only wise men of the world are philosophers."[31]

One of the most noteworthy of the condemnations was, of course, Article 34. In that item, Tempier condemned as erroneous the assertion that "the first cause [God] could not make several worlds." That prohibition clearly contradicted both Aristotle and most of the thirteenth century's leading scholars. As Duhem argued, Article 34 opened the door to a new freedom for scientists to explore possibilities about the cosmos beyond those that Aristotle permitted. More than once in his writings, Duhem observed 1277 as the year of modern science's origin.

"If we must assign a date for the birth of modern science, we would, without doubt, choose the year 1277 when the bishop of

Paris solemnly proclaimed that several worlds could exist, and that the whole of the heavens could, without contradiction, be moved with a rectilinear motion," Duhem wrote.[32]

Duhem believed that Tempier's proclamation freed philosophers to pursue science with an open mind, unencumbered by Aristotle's restrictions.[33] But later historians disagreed. Some suggested that medieval scholars weren't really encumbered by Aristotle all that much to begin with. Others simply disagreed on what constituted the birth of modern science. Such arguments hinge, of course, on what you mean by "birth" and "modern science." Maybe modern science was really born with Copernicus, or maybe it had previously been a small child and Copernicus was the kindergarten teacher. Did Galileo originate modern science? Maybe that was science completing undergraduate school.

In any case, however dubious it is to regard 1277 as science's birth year, the bishop's condemnation certainly did have a dramatic effect on the philosophical debate over the plurality of worlds. "It is a fact that the questions which Duhem thought so monumental were indeed widely discussed in the fourteenth century," commented Edward Grant, a noted scholar of medieval science. "Under penalty of excommunication, many deterministic arguments drawn from, or based on, Aristotle's philosophy had of necessity to be modified and qualified. Alternatives, previously thought to be silly or absurd, had now to be entertained as at least possible—even if only by virtue of God's infinite and absolute power."[34] And as Lindberg noted, "there can be no doubt . . . that the condemnations encouraged scholars to explore non-Aristotelian physical and cosmological alternatives."[35]

In any event, scholars commenting on the plurality of worlds after 1277 abruptly changed their tunes. With excommunication as a penalty for maintaining their previous beliefs, most suddenly realized that, after thinking about it, maybe God could make multiple universes after all.

Condemnation Aftermath

The general emphasis on God's absolute power that was fostered by
the Condemnation of 1277 encouraged many not only to take
seriously the possibility that God could create other worlds
than our own but to assume that he had actually done so.

—EDWARD GRANT, *PLANETS, STARS, & ORBS,* 1994

MENTION THE PHRASE "medieval times," and for most people you'll bring to mind images of knights in armor, sieges of castles, magnificent cathedrals, or possibly an amusement theme park. Science is unlikely to be anywhere in the picture. A few knowledgeable respondents might envision monks transcribing scientific manuscripts. But for the most part, the millennium between the fall of Rome in the fifth century to the beginnings of the Renaissance—an epoch often designated as the Dark Ages—is commonly viewed as a time when science was absent from the human agenda. In his famous TV show and book *Cosmos,* Carl Sagan reinforced that picture, alleging that "the sciences of classical antiquity had been silenced" for a millennium representing "a poignant lost opportunity for the human species."[1]

Sagan was apparently unfamiliar with Nicole Oresme.

In the midst of the Middle Ages, Oresme shone as a stellar example of medieval scientific scholarship. He was, wrote science historian Stefan Kirschner, without doubt "one of the most eminent scholastic philosophers," whose work foreshadowed much of modern science and mathematics. Oresme was also "generally considered the greatest medieval economist."[2] Historian Marshall Clagett noted that Oresme had been credited with "the framing of Gresham's law before Gresham, with the invention of analytic geometry before Descartes, with propounding structural theories of compounds before the nineteenth-century organic chemists, with discovering the law of free fall before Galileo, and with advocating the rotation of the earth before Copernicus."[3] Actually, Clagett added, those claims were incorrect, or at least exaggerated. But they all derived from serious scientific and mathematical investigations that Oresme had in fact undertaken with distinctive expertise.

Oresme's role as one of the most prominent medieval scholastic philosophers has often been exploited by historians to analyze the state of fourteenth-century scholarship. In particular, Oresme's writings illuminate the status of cosmological philosophy in the aftermath of Étienne Tempier's condemnation of 1277. Even decades after the condemnation, Oresme, among others, still cited Tempier's edicts in discussing the major issues facing medieval natural philosophers, including, of course, the question of the plurality of worlds.

MUNDUS ET UNIVERSUM

In the immediate aftermath of the condemnation of 1277, scholars quickly abandoned their absolute support for Aristotle's insistence on only one universe. "The decree of 1277," Pierre Duhem wrote, "marks a complete reversal in the opinion of the Parisian masters with respect to the plurality of worlds." Before 1277 scholars almost universally agreed with Aristotle—there could be only one world—and their writings forged elaborate chains of reasoning to demonstrate why Aristotle was right. But the threat of excommunication broke those chains. In the wake of Tempier's edict, philosophers scrambled to demonstrate the fallacies in Aristotle's reasoning, rather than its validity; "all theologians took it for certain that God could, if he wished, create multiple worlds."[4] The Franciscan Richard of Middleton, for instance, began defending the possibility of multiple universes, explicitly citing "the opinion of Lord Stephen, bishop of Paris . . . who excommunicated those who dogmatized that God cannot make more worlds."[5]

Soon philosophers at Paris and other centers of European learning allowed their minds to wander into the expansive intellectual space that a multiplicity of universes offered. One result was more elaborate attention to the nature of the question. Under the new theological rules, it was important to specify the terms of the debate more clearly. If you wanted to demonstrate logically why God could make other worlds, you needed to examine the meaning of the word *world* itself.

Obviously, *world* was not used by medieval scholars in the modern sense of a single celestial body. For them, the words *world* and *universe* (in Latin, *mundus* and *universum*) were essentially synonyms. Both implied the inclusion of all that existed. Earth was the central sphere of the medieval "world," one sphere among many others that carried the sun, moon, planets, and stars. Encompassing all of them was the outermost sphere, rotated by Aristotle's "Prime Mover," identified in the Middle Ages as the Christian God. For followers of Aristotle, that set of concentric spheres was the whole shebang—nothing outside of them existed, by virtue of the definitions of *exist* and *nothing*.

In the Aristotelian view of the cosmos, the Earth was not a celestial body; such bodies weren't even made of the same stuff. *World* usually did not refer to celestial objects until well after Copernicus, in the sixteenth century, showed that the Earth itself was just another celestial body. After that the question about the possible plurality of worlds had to be continually reformulated because the conception of the cosmos, or universe, kept changing. Much of the story of the multiverse debate throughout history concerns this continual revision of the concept of *universe*.

As debates about multiple worlds became more sophisticated, medieval philosophers realized that the term *world* could be taken in different senses. One scholar who addressed the question toward the end of the thirteenth century was William of Ware (or William Varon), a Franciscan who studied at Oxford and who taught there and also at Paris.[6]

Varon was known as Doctor Fundatus and also sometimes as Doctor Praeclarus, for being "very clear," which he more or less was in discussing the nuances of the term *world*.[7] "By world one can understand the universality of creatures taken as a whole," he wrote. "Then a world other than this one, once created, would not be constituted of the universality of creatures. It would therefore not be another world, but only one part of the universe."[8] On the other hand, "another world" could refer to another set of celestial spheres. In such a world, the four elements would be arranged just as the four elements are ordered in our world. It's this second sense of *world* that should be considered in asking if God could make more than one, Varon pointed out.

Later thinkers, such as Thomas of Strasbourg (c. 1300–1357), sharpened the question. He articulated a concern that still infuses the multiverse debate today—namely whether it makes any sense to speak of the universe as anything other than everything that exists. In the present day the Nobel laureate Frank Wilczek has dealt with just this issue by defining the "universe" as "the domain of physical phenomena that either are, or can reasonably be expected to be, accessible to observation by human beings in the foreseeable future."[9] That "universe" could be part of a grander structure of other out-of-sight universes making a multiverse.

Thomas of Strasbourg defined universe somewhat differently, conceiving it as "a unity in which all things are enclosed" (in the Aristotelian universe, enclosed by the outermost sphere). But if some other "world" existed outside that sphere, the totality would no longer be a unity, defying the definition of universe. In other words, the idea of more than one universe contradicted the definition of *universe*. That made multiple universes an impossibility, even for God, because even he could not do anything that implied a contradiction. So Thomas suggested a second definition of *universe* or *world:* it could apply to "any collection of things, whether identical to or different from the things in our world."[10] Creating another collection of things of whatever sort was certainly within God's power.

WORLDS WHOSE MULTITUDE IS POTENTIALLY INFINITE

Among the scholars affirming God's power to make many worlds was the prominent theologian Henry of Ghent (d. 1293), nicknamed Doctor Solemnis. He was among the theologians convened by Bishop Tempier to prepare the list of condemned teachings in 1277. Modern research suggests that Henry was a freethinker who did not actually agree with all of the condemnations on Tempier's list.[11] But Henry did support God's power to make multiple worlds. "God can create a body or another world beyond the ultimate heaven," Henry wrote, just as God made our world in the first place.[12]

Another Parisian theologian, Godfrey of Fontaines (a contemporary of Henry's), disputed the pre-1277 argument that there could be no other world because all the matter that exists had been used to make this one. God produced *that* matter, Godfrey replied, so if he wanted to he could produce more matter to make another Earth beyond the spheres of our universe. "The existence of another world is not therefore rendered impossible," Godfrey wrote.[13] At Oxford, William Varon made the same point: if God made matter from nothing once, he could do it again. If God created a man containing all human matter, it would not prevent God from creating another man. He wouldn't do it by taking some of the human matter from the first man, but rather by making a new one from scratch.

Varon also denied that another world would confuse the element earth about where to go, disputing the contentions of Albert the Great and Thomas Aquinas. Varon declared that another universe would be a "separate domicile" for that element: earth within our sphere knows its natural place, and earth in another world, outside our sphere, would not seek the center of our world. Rather it would reside in its own world's center. Varon also vigorously contested the notion that lack of a vacuum could have prevented God from creating more worlds. "Before the creation of our world there was nothing here, and God created this world; thus He can create another world outside our world. . . . He can create, where there is absolutely nothing, other worlds whose multitude is potentially infinite."[14]

Varon's main point is that no matter how many worlds God created, he could always still create more. The known world, which God made, is finite; creating it could not have exhausted his power, which is infinite. A father's power to produce a son is not exhausted by producing a son, Varon pointed out—it remains possible to produce more sons.

Other late thirteenth-century thinkers devised arguments favoring the possibility of multiple worlds—mainly to establish that their teachings resided safely within Bishop Tempier's restrictions. Few scholars seriously proposed that such other worlds actually existed. It was just important to show that Aristotle's reasoning (and the reasoning of pre-1277 natural philosophers such as Bacon and Aquinas) had flaws. With those flaws corrected, God was free to create more universes without the need to circumvent logical contradictions. In almost all cases these corrections boiled down to exploiting God's supernatural powers to do as he pleased without regard to natural processes. That is to say, other universes were miracles, deviations from the natural order of things. Another universe, separate from the known world, would be a continuous permanent miracle. That would be perfectly within God's power, if he chose to so exercise it. But it was probably not anything God would actually bother doing, most scholars surmised. Albert of Saxony (d. 1390), for instance, defended Aristotle's arguments against multiple worlds, concluding that more than one world was impossible, "naturally." "It is no less true that God could create many worlds, since He is omnipotent," Albert added, to acquiesce in Tempier's edict.[15]

In the fourteenth century, though, some scholars explored the multiple universe idea a bit more seriously. They seemed not just to be protecting themselves from excommunication, but in fact believed that the possibility of other worlds was worthy of genuine contemplation.

BURIDAN'S ASSESSMENT

In Paris, the preeminent natural philosopher of the early fourteenth century was John (in French, Jean) Buridan (1300–1358), most famous in popular lore for his ass. Nothing obscene about it: his ass was a hungry donkey, or jackass, positioned exactly halfway between two mounds of hay (or in some versions, a hungry, thirsty donkey placed midway between a mound of hay and a pail of water). It was an illustration posed by later philosophers to describe Buridan's views on making the tough choice between two equally desired alternatives.

Using the standard format of posing questions (based on Aristotle's *De Caelo*) and then analyzing them, Buridan asked "whether it is possible that other worlds exist." He began by summarizing the arguments in favor of a "yes" answer provided by Aristotle (who always presented others' arguments carefully before disputing them). Those "yes" points included such grammatical considerations as "world" being singular, which implies the possibility of a plural. And if God could make one world, which he did, he could surely make another. "He is not of less power now than He was then."[16] And on top of all that, the world is good, and since more than one good thing is better than one good thing, it is at least possible that more worlds should exist in order to achieve a greater good.

Buridan then cited Aristotle's view that multiple worlds posed a logical contradiction because the element earth would be impelled to move naturally in different directions. Buridan also noted that Aristotle believed multiple worlds implied multiple gods, because one god is simple and unique and therefore would not make different worlds.

A subtlety in the argument, Buridan pointed out, is that "plurality of worlds" can be imagined in two ways: multiple worlds might exist simultaneously or successively. That is, they could exist outside of this world, or they could precede or follow this world in time.[17] God could make different worlds one after the other in time (since he had the power not only to create, but also destroy such worlds). But such construction and destruction can occur only by divine action. To the argument that the world changes daily, and is therefore a succession of different worlds, Buridan cited the counterargument that the world stays the same in its most important major parts, so it is correct to maintain that the world remains the same "from day to day."

In summing up and presenting his conclusions, Buridan dismissed the grammatical arguments. And he insisted that more gods or worlds are not justified as possible on the grounds that they would be better, because in some ways they would be worse, and being both better and worse is a logical contradiction. So that proposition, being contradictory, is impossible. But to the point that since God created this world he could create others as well, Buridan assented. "We hold from faith that just as God made this world, so he could make another or several worlds."[18] In this way Buridan established his compliance with Article 34 of the Condemnation of 1277.

Still, Buridan's concession that God could create more worlds does not appear to signify a belief that he had actually done so. Buridan elsewhere stressed that even if God could create another world, he would have no reason to, since if he wanted more creatures it would be simpler and more economical just to make this world bigger. Creating a new space outside our world would therefore be without necessary purpose.[19] That sounds a little bit like Buridan was invoking William of Ockham's razor. But William, Buridan's contemporary, reasoned in basically the opposite direction.

OCKHAM'S RAZOR

Ockham, a notoriously nonconformist scholar, was born in Ockham (not a coincidence) sometime in the late thirteenth century, exact date unknown, but probably before 1290.[20] He studied at Oxford and became a Franciscan friar. Apparently he fulfilled the requirements for his theology degree, but it was never awarded, seemingly blocked by an intellectual enemy. (Academic politics of an unsavory nature were not invented in the twentieth century, after all.) His nickname, Venerabilis inceptor, reflected his lack of degree, as inceptor means something like "beginner."

Ockham's outside-the-box views led to complaints of heresy (engineered by his academic enemy), and he was called to Avignon in France (home of the pope in those days) to defend himself. While the investigations related to the charges took years, Ockham's political maneuverings eventually put him in league with a few other renegades of the Franciscan order, and they ended up supporting Holy Roman Emperor Louis IV in Germany, who had installed an anti-pope. Ockham lived the last two decades of his life in Munich, where he died in the late 1340s, possibly from the Black Death.

Before all this trouble, though, Ockham had established himself as one of Europe's prime natural philosophers and logicians, "perhaps the greatest logician of the Middle Ages."[21] He had been influenced by the Scottish philosopher John Duns Scotus (c. 1266–1308) but developed his own views, often at odds with prevailing prejudices. For Ockham, reason was a gift from God, so it was necessary to use it properly. And Ockham understood that logic alone is not sufficient to guide reason. Experience and observation provide the material necessary for logic to analyze. "We have to find out how things are, we have no right to dictate how things must be," Ockham held.[22] A good rule to live by for many professions or walks of life.

Ockham is most famous, of course, for his razor. He was not the first to articulate it, though, and he never used the term *razor*. And he never said it in the way that it is sometimes popularly expressed today—that the "simplest" explanation is the correct one. But the basic idea of the explanatory power of parsimony flowed through his work. "Plurality is not to be posited without necessity" was one of his phrasings of it; another was "what can be explained by the assumption of fewer things is vainly explained by the assumption of more things."[23] Other translations give "it is vain to do with more what can be done with less" and "it is futile to accomplish with a greater number of things that which can be accomplished with fewer."[24] (A phrasing often attributed to Ockham, "entities must not be multiplied without necessity," doesn't appear in his writings.)

Ockham's principle was expressed similarly in Isaac Newton's first rule for reasoning in natural philosophy: "We are to admit no more causes of natural things than such as are both true and sufficient to explain their appearances." As the philosophers say, Newton continued, "Nature does nothing in vain, and more is in vain when less will serve."[25]

Ockham's razor is very often misunderstood. Its popular version expressed as "the simplest explanation is best" or even "the simplest explanation is right" is vastly oversimplistic. These phrasings of Ockham's razor illustrate how easily it is misinterpreted, as it often is, even by scientists. (I recall one conference where a scientist-theologian argued that saying the universe was created by God was the simplest explanation and therefore the best.) In many modern arguments, Ockham's razor is often mistakenly evoked as though it were a law of nature, rather than just a Pirates-of-the-Caribbean sort of "guideline" for sound thinking.

In particular Ockham's razor is frequently called on by some who argue against the scientific validity of the multiverse. Cosmologist George Ellis has written, for instance, that "we are supposing the existence of a huge number—perhaps even an infinity—of unobservable entities to explain just one existing universe. It hardly fits . . . Ockham's stricture that 'entities must not be multiplied beyond necessity.'"[26] Paul Davies, a cosmologist well-known for his numerous popular books, calls the multiverse a "flagrant breach of the most basic rule of science, which is Occam's razor." (At least Davies acknowledges that that's a "knee-jerk reaction.")[27] As we will see in chapter 14, using Ockham's razor to refute the multiverse is a flagrant breach of Ockham's meaning.

Ockham, of course, cannot be held responsible for modern-day misreadings of his philosophy. And in any case, he did not have a problem with the multiverse. Among the scholars of his day, he seemed to embrace the possibility of multiple worlds with the most enthusiasm.

For one thing, Ockham adapted Richard of Middleton's anti-Aristotelian view that earth placed in another world would have no trouble figuring out where to go. Ockham reconstructed that argument using an analogy with fire, another of the four elements. Fire naturally went up, the opposite of earth's natural motion downward. If you set a fire in Oxford, Ockham wrote, it would fly up to the point on the heavens directly above Oxford. Fire in Paris would rise straight up above Paris. But if you somehow transported the Oxford fire to Paris, the Oxfordian flames would rise directly upward from Paris, not seek a circuitous route toward the heavenly point above Oxford. Natural motion does not, therefore, require motion toward only one single destination. Earth (the element) in another cosmos would naturally move toward the center of that cosmos, no matter where that earth came from to begin with, just as fire flies straight up regardless of its point of origin.

"If someone were to take the earth belonging to the other world and put it inside our heaven, it would tend toward the same place as our earth. But when it is outside this world, when it is inside the other heaven, it would no longer move toward the center of our world . . . but it would move toward the center of the other world," Ockham asserted.[28] In this way Ockham argued that God could make another world without logical contradiction. It was kind of an effort to make Aristotle consistent with Aristotle.

Aristotle's defenders would have disagreed. Motion can only be either natural or violent, and if natural motion (say, of earth in another universe) takes earth away from some point (center of our universe), then it can get closer to that point only by violent motion. Motion toward the center of our cosmos, then, could be both violent and natural, a contradiction.

Ockham responded by calling attention to a point that is still very important for science today—the role of initial conditions. If matter moves farther from some point under all circumstances, it's true that moving closer to it would require violent motion. But in the examples Ockham had articulated, earth moves away from our universe's center only for a certain set of initial positions. Discerning natural from violent motion requires taking initial conditions into account. In modern terms, the laws of nature by themselves do not always tell you what will happen to a system; you need to know the laws, but you also need to know what initial conditions to apply the laws to.

Ockham therefore argued strongly that multiple worlds, if not at present a reality, are undoubtedly a possibility. He framed his ideas in the form of commentary on the *Sentences* of Peter Lombard, the standard theological text of the time, which provided the basis for much original medieval scholarship. "I say that God can make a better world than this one," Ockham wrote.[29] God had the power to make an infinity of individuals, many more than exist now or have existed, but there is no reason he would have to make those individuals in this world. "He could produce them outside of this world and make [another] world of them, just as He made this world."[30] Whereas Buridan implied the need to make more individuals could be accommodated just by a bigger world, Ockham concluded that the possibility of more individuals implied the possibility of more worlds.

Ockham's writings clearly demonstrated that the 1277 condemnation had opened the way for a thorough discussion of the plurality of worlds. No such discussion was more thorough, though, than that provided by Nicole Oresme.

IMMENSE GENIUS

Of the great thinkers of the fourteenth century, the most impressive, wrote mathematician James Franklin, was Nicole Oresme. "Before the disaster of the Black Death in 1348–49, there appeared two men of immense genius," Franklin wrote.[31] One was Nicholas of Autrecourt.[32] One was Nicole Oresme.

They were not as influential as they deserved to be, though. The Black Death killed many of their would-be readers and followers. And "their writings were simply too profound to be understood by the intellectual pygmies by whom they had the misfortune to be succeeded," Franklin opined. Oresme, though, was eventually read avidly by modern historians. As he was the preeminent scholar of his era, his works provided a comprehensive picture of the state of fourteenth-century natural philosophy. His work is especially noteworthy for demonstrating that medieval scholars pre-Copernicus were not, as popularly viewed, all slavish adherents to whatever Aristotle believed. "Since Oresme was the leading savant of his age, his criticism of the Aristotelian system may fairly be considered to represent the most advanced thought of the time," wrote Oresme scholar Albert Menut.[33]

Born in Normandy about 1320, Oresme was educated at the University of Paris, where he was no doubt an outstanding student. He may or may not have been taught by Buridan but was certainly familiar with Buridan's writings as well as those of Ockham, whose influence on natural philosophers was widespread during Oresme's youth. By sometime in the 1350s Oresme had earned his theology doctorate. After a period of teaching at the university, he turned to serving government and the church. Eventually he became the Bishop of Lisieux in 1377, five years before he died.[34]

Oresme's greatest body of work was in mathematics. He was a fan of Grosseteste's view that math was essential to sound knowledge. Oresme applied his mathematical expertise to everything from the geometric laws of optics to the description of the heavens (and even economics). He devoted much of his energy to understanding motion, one of the prime problems of medieval natural philosophy.

While teaching at Paris, Oresme had the good sense to make friends with the dauphin of France, who would later become King Charles V. It was a fortunate friendship not only for Oresme but also for science, as Charles appreciated intellect and supported Oresme's career. In fact, it was at the king's request that late in life Oresme undertook a translation of Aristotle's *De Caelo* (full title, *De Caelo et mundo*), which appeared in French as *Le Livre du Ciel et du Monde* (*The Book of the Heavens and the World*). In *du Ciel,* Oresme applied the accumulation of his wisdom to interpreting (and correcting) Aristotle's analysis of the cosmos. In stating his purpose for the translation, Oresme made clear that his goal (shared with Charles) was not only doing science (or natural philosophy) but also spreading the knowledge gained to the outside educated world. Hence the translation, and commentary on it, appeared in the vernacular Middle French. For a text to translate, Oresme used a Latin version that had been translated from Greek by Robert Grosseteste and William of Moerbeke (who completed the translation after Grosseteste's death).

In those days, translators generally did not stop with rendering the original in a new language. The point was not only to translate but also to comment. Oresme constantly interrupted his translation with either short comments, summarizing or explaining what

Aristotle had just said, or longer digressions expositing Oresme's own ideas. As a result the modern English translation of Oresme's French version is a nice fat book of more than 700 pages, with Middle French on each left-hand page and the English translation on the right. Within those pages is Oresme's incisive analysis of the question of the plurality of worlds.

There is, of course, much more than that. In this book Oresme first suggested the metaphor for the universe as a clockwork mechanism, a notion that later became identified with the Newtonian worldview of the cosmos. Oresme also produced a fascinating demonstration that the Earth might very well spin on its axis, suggesting that the nighttime sky remained stationary as the Earth rotated beneath it, rather than the other way around. In the end he concluded that the Earth did, in fact, stay still, as scripture required. His point, though, stood: there was no logical physical reason why that must be the case.

An important part of Oresme's writing more closely related to multiple worlds was the possible existence of a vacuum beyond the outermost sphere of the Aristotelian cosmos. That issue occupied many of the great minds of the Middle Ages, most concurring with Aristotle that such an external vacuum was not possible. Their arguments boiled down to a Gertrude Stein–like conclusion that there simply was no there there. A body's "place," as construed by Aristotle, was defined as the inner surface of the object containing the body. Outside the outermost sphere, there was nothing to contain anything, therefore there could be no place, and no body could exist there. And so there could be no other worlds beyond the known world of Aristotelian spheres.

INFINITE IN HIS IMMENSITY

Until 1277 the question of a "void space" beyond the cosmos was generally dismissed quickly, as Aristotle had supposedly proved it impossible. After 1277 scholars recognized the necessity to allow such a space under the "God can do anything" clause. And Article

49 of the 1277 condemnation explicitly prohibited teaching that God could not move the world from one place to another (that is, by rectilinear motion) and that he was unable to do so because a vacuum would remain in the vacated space. That edict amounted to the assertion that a vacuum beyond the cosmos could in fact exist. (Article 201, on the other hand, implied that no vacuum existed before God created the world.) Still, few scholars maintained the vacuum's existence to be likely. Even among those who took the question seriously, most believed that an extracosmic void did not make sense—especially, as some considered, if that space would be infinite.

Buridan, for instance, denounced the idea as unreasonable. "We ought not to posit things that are not apparent to us by sense, experience, natural reason, or by the authority of Sacred Scripture," he wrote. "In none of these ways does it appear to us that there is an infinite space beyond the world."[35] Except for the sacred scripture part, these arguments sound very much like the position of many today who oppose the modern idea of the multiverse. If we cannot observe these other universes, such reasoning goes, they are out of the bounds of the scientific playing field. Throughout history, though, such arguments have not prevented some thinkers from exploring those out-of-bounds regions. Like Oresme.

In the fourteenth century, he was one of the few who argued not just for the possibility of a void beyond the spheres, but also advocated strongly for its existence. It was clearly possible to imagine such a space, he said. In fact, the extracosmic vacuum was often referred to as "imaginary" space in those days. But the meaning of "imaginary" space differed from scholar to scholar.[36] Some construed it to mean that the extracosmic void was a figment of the imagination, a "mental fiction." But a few regarded it as space inaccessible to human senses that could only be imagined, while nevertheless really existing. Oresme was one who clearly believed that not only could a vacuum exist beyond the last sphere, but that it also did in fact exist and extended to infinity. This immensity of space, Oresme declared, should be equated with God (thereby perhaps avoiding conflict with Article 201).

It was in this context that Oresme explored the notion of a plurality of worlds. By insisting that space existed outside the cosmos, Oresme did away with the common argument that no external world could exist because it would require a vacuum. There *is* a vacuum, so that objection fails. Oresme similarly dismissed the old argument that all the matter had been used up to make this world: "In truth, God could create *ex nihilo* new matter and make another world." As for the confusion caused if earth from another world found itself in this one, Oresme saw no problem. "If some part of the earth in the other world were in this world, it would tend to the center of this world," he wrote. But such parts of earth in their own world would not try to get to this world—"in their world they would form a single mass possessed of a single place," just like the earth in our world. Aristotle's argument that other worlds imply the need for more than one god, when there can be only one, similarly fails: "God is infinite in His immensity, and, if several worlds existed, no one of them would be outside Him nor outside His power," as God existed in (or was identical to) all the infinite space. "I conclude," Oresme insisted, "that God can and could in His omnipotence make another world besides this one or several like or unlike it. Nor will Aristotle or anyone else be able to prove completely the contrary."[37]

Oresme did not leave it at that. He explored several imaginative possibilities for what it would really mean for other worlds to exist. When discussing the possible plurality of worlds, he declared, it was important to do so "without considering the authority of any human but only that of pure reason." And reason offers three ways to imagine other worlds (one more than the two mentioned by Buridan). One, as Buridan and others had identified before, is the possible existence of multiple worlds succeeding each other in time. As some ancient thinkers had imagined, the present world was formed by chaos, "love or concord" having disentangled the mess and created a new world with order. Conceivably that world might someday descend back into chaos, returning to "the same confused mass." And "through concord, another world will then be made."[38]

In principle this creation-destruction cycle would repeat itself eternally. (A similar view of a cyclic universe had been developed by the Stoic school of philosophy in ancient Greece and Rome, and in modern form has been revived under the label of "ekpyrotic" cosmology.) Oresme seemed to think that such an infinite succession of universes would not happen naturally, although God could make it happen if he wanted to.

Oresme took special interest in a second type of other world—one hiding inside another. For all we know, our set of spheres is just nested within a larger set of spheres, and deep below our feet exists another miniature set of spheres, its inhabitants similarly unaware of the greater cosmos surrounding it. Oresme emphasized that this idea "is not in fact the case, nor is it at all likely." But it is illuminating, he says, "to toy with as a mental exercise."[39]

His point is that examining this idea uncovers no way to prove it impossible by logical argument. Large and small are relative terms. A world concealed within ours would seem to anyone there just the same as ours does to us; an even larger world would look the same to its inhabitants. "Were the world to be made between now and tomorrow 100, or 1,000 times larger or smaller than it is at present, all its parts being enlarged or diminished proportionally, everything would appear tomorrow exactly as now, just as though nothing had been changed," Oresme asserted. After analyzing various objections to the nested multiple worlds, he concluded that neither reason nor evidence from experience can prove the impossibility of such an arrangement. On the other hand, there is also no proof from reason or experience that such worlds actually do exist. It would be unwise to contend that such worlds exist without any reason or cause for so believing. But it was a good idea, he said, "to have considered whether such an opinion is impossible."[40]

Finally, Oresme took up the third (and most often considered) notion of a plurality of worlds: those existing outside of the space occupied by the known cosmos. Aristotle rejected that idea, of course, but Oresme held that Aristotle's arguments "are not clearly conclusive." After describing and rebutting Aristotle's reasons for denying such worlds, Oresme argued that Aristotle's error lies in not recognizing the special nature of the space beyond his cosmos, the infinite space that Oresme identifies with God. "Outside the heavens, there is an empty incorporeal space quite different from any other plenum or corporeal space," Oresme declared, infinite in spatial extent and eternal in time. That infinite space differs from ordinary space just as eternity differs from ordinary duration. An infinite eternal universe exists, Oresme proclaimed, because it is God himself: "This space of which we are talking is infinite and indivisible, and is the immensity of God and God Himself, just as the duration of God called eternity is infinite, indivisible, and God Himself."[41] We cannot comprehend this eternal space, Oresme wrote. Nonetheless "reason teaches us that it does exist."

And so God could, in fact, make multiple worlds. Plenty of real estate was available. But did he? Out of the blue, Oresme suddenly says No! "But, of course, there has never been nor will there be more than one corporeal world," he insisted. For Oresme, theological and scriptural considerations (there was no mention of other worlds in the book of Genesis) outweighed the logical possibility of many universes.[42] He considered them instructive fodder for "mental exercises," not physical realities. Nevertheless, Oresme's analysis demolished virtually every argument against multiple worlds, showing them to be a possibility in principle. And for one of the leading thinkers of the following century—Nicholas of Cusa—that possibility seemed undeniably real.

Cusa and Copernicus

We should therefore learn . . . not to look on our earth as the universe of God, but as a single, insignificant atom of it; that it is only one of the many mansions which the Supreme Being has created for the accommodation of his worshippers.

—E. H. BURRITT, *THE GEOGRAPHY OF THE HEAVENS,* 1833

WHEN HE REMOVED the Earth from the center of the universe and replaced it with the sun, in one sense Nicolaus Copernicus didn't think it was such a big deal.

After all, the sun and Earth really weren't so very far from each other. But both were unimaginably far from the sphere carrying the fixed stars—the shell surrounding the entire cosmos. Since the Earth resides relatively close to the true center (occupied by the sun), and the universe is so vast, humankind's viewpoint still seems to be pretty much from the middle.

"It is certainly clear enough that the heavens are immense in comparison with the Earth and present the aspect of an infinite magnitude," Copernicus wrote in his masterpiece, published in 1543. So even though the Earth isn't really precisely in the middle of it all, its displacement from the center is minuscule. "Although it is not at the center of the world, nevertheless the distance is as nothing," Copernicus wrote, "particularly in comparison with the sphere of the fixed stars."[1]

Of course, in many other respects it was a big deal, as Copernicus well knew. He did revolutionize astronomy, after all, and he pretty much demolished the underlying premises of Aristotle's philosophy of matter and motion. In terms of the physical construction of the cosmos, though, Copernicus's was not all that terribly different from Aristotle's—just a little reshuffling of celestial bodies in its core.

But a century before Copernicus completed his rearrangement of the cosmos, a revolutionary and imaginative thinker had prescribed an even more radical reconsideration of cosmic architecture. For Nicholas of Cusa, the question of whether the Earth or sun occupied the center of the universe made no sense—because in his view there was no such center to begin with.

Though vastly different in many respects, Cusa's and Copernicus's new views of the cosmos shared one noteworthy import: they both radically altered the entire question of what constituted a multiplicity of worlds.

LEARNED IGNORANCE

Imagine a circle with an infinite diameter. Take a minute if you need to. Just envision a circle getting bigger and bigger, until you can see only a portion of its arc in your mind's eye. Now it's just a curved line, getting less curved as the circle gets bigger. As you continue to mentally expand the circle, your mind sees the arc's curvature constantly diminishing. Soon only a small portion of the circle will be visible, and eventually it will look like a straight line. It's just like why the Earth, in fact a large sphere, looks pretty darn flat to anybody standing on it.

But the Earth's surface really is a little bit curved, as the planet is finite. For an infinitely large circle, the line tracing its circumference would in fact be absolutely straight. It would extend forever in whichever direction you chose to explore it, from whatever point you started from. There is no point on the line that is any different from any other. No center, or midpoint, of the line exists. Nicholas of Cusa, who imagined such a circle six centuries ago, concluded from this simple exercise that the universe possesses no special place—there is, in fact, no center of it for the Earth to occupy.

In a universe with no center, Aristotle's prime argument against a plurality of worlds collapses. He argued that earth (the element) sought its natural place in the center of the cosmos. Two universes would mean two centers, depriving earth of a unique natural resting place, creating what Aristotle considered a logical contradiction. But if there is no center of the universe, then Aristotle's logic is moot. An infinite universe could contain an infinity of worlds. Cusa embraced the idea of many worlds, populated, he speculated, by intelligent beings of various types, suited to their environment, which differed from world to world.

No doubt one of the most original thinkers of all time, Cusa perceived simple (to him) truths and from them drew grand deductions. And he succeeded far more than his predecessors in escaping Aristotle's powerful grip on European intellect. "Trained in the methods of Aristotelian logic, Nicholas found them inadequate to his purpose," wrote one commentator—that purpose being to put "mathematics and experimental science at the service of philosophy."[2]

Cusa was born about 1401 in Kues (hence Cusa) on the Moselle River in Germany. He attended the University of Heidelberg and later continued his education at the University of Padua, where he studied church law, earning his doctoral degree in 1423. At Padua, he heard Prosdocimo Beldomandi lecture on astrology and attended sermons preached by Bernardino of Siena, a prominent Franciscan. Later Cusa taught in Cologne and researched church law there, making some noteworthy discoveries about legal history and turning up some important (but supposedly lost) Latin scientific writings, such as Pliny the Elder's *Natural History*. In

1430 Cusa became an ordained priest and subsequently took part in various church-related legal proceedings and diplomatic missions.[3]

During this time Cusa had thought deeply about the philosophy of human knowledge. Familiar with the mathematical writings of some earlier scholars, such as Thomas Bradwardine (c. 1290–1349), Cusa attempted to conceive the limits of human knowledge in the framework of mathematical truths plus the application of Aristotelian logic. Aristotle had a severely insufficient grasp of infinity, Cusa concluded, spelling out his case in a major treatise titled *De Docta Ignorantia—On* (or *Of*) *Learned Ignorance.*

In this treatise, Cusa outlined his insights into the nature of human knowledge and its relation to God—who, in his infinity, is forever beyond complete human comprehension. Seeking knowledge, Cusa contended, is in essence seeking, in modern political language, "known unknowns." "Assuredly," he wrote, "we desire to know that we do not know. If we can fully attain unto this [knowledge of our ignorance], we will attain unto learned ignorance. . . . And the more deeply we are instructed in this ignorance, the closer we approach to truth." But we can only approach the truth more and more closely; never can we attain truth in all its purity. "Though it is sought by all philosophers, it is found by no one as it is."[4]

Cusa was no cosmologist. His treatise is all about understanding God, not the universe. But in formulating his view of God, Cusa derived implications for the nature of the cosmos and Earth's place in it. At the foundation of his approach is an obsession with the concepts of maximum and minimum. For Cusa, maximality indicates that of which nothing can be greater. Obviously the maximum must contain everything else, including the minimum, of which nothing can be lesser. God is an "Absolute Maximum"; the universe is a limited or "contracted" maximum, not quite infinite because not as great as God.

In Cusa's view, all things in the cosmos are "unfolded" from God. So our knowledge that all things unfold from God—the

Absolute Maximum—enables us to deduce many things about the world, Cusa believed.

He built much of his philosophy on his contention that the maximum is in some way equivalent to the minimum. Since the absolute maximum is all that can be—and is, in actuality, all that is—it cannot be less than what it is, which is everything. So it cannot be lesser. At the same time the minimum is that of which there cannot be lesser. Since both the maximum and minimum cannot be less than they are, Cusa deduces that the minimum coincides with the maximum.

If this reasoning seems a little sketchy, don't worry—it was hard to grasp by readers in Cusa's day, too (which is perhaps why his work had little impact during his lifetime). Scholars have for centuries complained about the density of Cusa's language. "That he failed always to be understood and appreciated by his contemporaries is due in part to the peculiarity of his language, neither medieval nor humanistic and flawed by inadequately defined words and concepts," one scholar has commented.[5]

Cusa authority Jasper Hopkins acknowledges that Nicholas "has not made his reader's task easy." Although Cusa says he endeavored to explain his views "as clearly as I could," nonetheless "many of his points escape even the diligent reader, since the explanation for them is either too condensed, or else too barbarously expressed, to be assuredly followed," Hopkins writes.[6] But it's worth the effort, Hopkins contends. You can't just read Cusa's works—you have to study them, and with enough diligence, a coherent view of God and the universe emerges. (So if the paragraphs that follow take a little extra effort, console yourself with the happy thought that you don't need to read Cusa's complete works in their entirety.)

It was in an attempt to aid his readers' comprehension that Cusa chose the illustration of his famous infinite circle. For an ordinary circle, its diameter (a straight line) is shorter than its circumference (a curved line). But as a circle grows, the diameter increases while the curvature of its circumference diminishes. Ultimately the circumference becomes maximally straight and minimally curved. So for such an infinite circle—a maximum circumference—its diameter is also at a maximum. There cannot be more than one maximum, Cusa pointed out, so the diameter is the same as the circumference.

And not only is the diameter equal to the circumference, but the middle of the diameter is also infinite—and the middle is the center. Thus the center, diameter, and circumference are all the same. From this reasoning Cusa concluded that the Absolute Maximum is not comprehensible (and on this point few would argue with him). That is the "learned ignorance" that humans must face: the incomprehensibility of the Absolute Maximum, that is, God.

Of course, the circle is just an illustration. In actuality, the universe (Cusa presumed) is spherical. But it's a simple matter to extend such reasoning about a circle to the spherical cosmos—you simply have to rotate the circle to generate a sphere. (Most of you probably don't remember watching Mr. Wizard demonstrate this on TV by putting a round disk on the end of an electric drill. But I do.) Remember, the universe is not the Absolute Maximum that is God, but a limited maximum, contracted from the Absolute Maximum (but imitating it). "In this respect . . . it is evident that the whole universe sprang into existence by a simple emanation of the contracted maximum from the Absolute Maximum," Cusa declared. Since the universe is contracted from God, God is in the universe; things within the universe are contracted from the universe, so the universe is in those things; therefore, God is in those things.[7]

WORLDS WITHOUT NUMBER

This presence of God in things explains why the world is as it is: not random as Epicurus held, but formed from the possibilities inherent in the Absolute Maximum. The universe and things within it—the Earth, the sun, and everything else—"sprang forth rationally from possibility" latently present in matter contracted from the Absolute Maximum. Thus the world and its contents were a rational expression of an "aptitude only for being this world" contained in the contracted possibility. "Unless the possibility of things were contracted, there could not be a reason for things but everything would happen by chance, as Epicurus falsely maintained," Cusa declared.[8]

Given this foundation, Cusa could then draw conclusions about the world quite contrary to the views of his medieval predecessors, not to mention Aristotle. Cusa explained that the universe cannot be conceived as finite, because it is not enclosed by any boundaries, yet it is not infinite (having been contracted from the Absolute Maximum). Drawing on the analogy of the infinite circle, he pointed out that the Earth cannot be the center of such a universe, for it is not possible for the cosmos to have a fixed center, whether occupied by earth or air or fire or anything else.

And just as the Earth is not at the center, the outermost sphere of fixed stars does not constitute the circumference of the cosmos (as there is no circumference). The apparent revolution of the sphere of stars over the Earth does not imply that the Earth is stationary at the center of that sphere (or any sphere); Earth, Cusa proclaimed, is just another "star," or celestial body, that moves.

But if the Earth moves, why does it seem to be standing still while the stars rotate around and the planets wander through the night sky? It's because, Cusa asserted, we discern motion only by comparison with something stationary. He uses the example of a boat on a flowing river: a person on that boat unable to see the shore would have no way of knowing that the boat was moving with the flow. Similar reasoning applies to inhabitants of any celestial body. Wherever you are—on the Earth, the sun, or even some other star—you would consider yourself immovable at the center of all the motions you saw in the sky. While he did not work out this idea rigorously, Cusa clearly had some concept of the relativity of motion that was later to be articulated more thoroughly by Galileo (who used his own, somewhat different, ship analogy).

From this analysis flowed Cusa's contention that the Earth is just a star, albeit a "noble" (or "brilliant") one, distinct and different from all other stars.[9] So for Cusa, the question of a plurality of worlds was posed in the context of a radically revised version of Aristotle's cosmic set of concentric spheres with Earth at the center. Now the worlds (or stars) were all like the Earth in some way, celestial bodies in motion, expelling heat and light, and possibly providing habitation for intelligent and other beings— "men, animals and plants."[10]

No previous medieval thinker had been quite as explicit in affirming the possible existence of intelligent life on other worlds, or in asserting most assuredly that numerous other worlds exist— so numerous, in fact, that they are "without number" (by which he meant an indefinite or uncountable number, not an infinity).[11]

Cusa conjectured that none of the other regions of the stars lack inhabitants; each region is "like the world we live in, a particular area of one universe which contains as many such areas as there are uncountable stars." In other words, there are as many places suitable for life in the universe as there are stars, which are too numerous to be counted. He pondered the possible relationship of these extraterrestrials to Earth-based life and supposed that they would be much different, exhibiting features suitable to their environment. "It may be conjectured that in the area of the sun there exist solar beings, bright and enlightened intellectual denizens, and by nature more spiritual than such as may inhabit the moon . . . whilst those on earth are more gross and material."[12]

For the plurality-of-worlds issue, the significance of Cusa's work rested largely on his redefining the nature of the debate. In Cusa's view, all celestial bodies—even the sun—were made of the same elements. Earth (the element) could no longer get confused by the existence of another world; the universe contained many celestial bodies, or centers of attraction, for earth to move toward. "It is likely that it was through the concept of a 'center of attraction' that the tradition of the plurality of worlds shifted its focus of attention from the plurality of Aristotelian *kosmoi* to the plurality of Earthlike planets," according to Steven Dick.[13]

Nobody paid much attention at the time to this implication of Cusa's thought, except perhaps for Leonardo da Vinci (1452–1519). Among Leonardo's multifaceted interests was the nature of the universe, and some scholars have argued that the influence of Cusa drove Leonardo away from adherence to Aristotle.[14] Leonardo found no problem with the idea of many worlds, adopting a belief similar to Cusa's about the existence of multiple centers of attraction. Leonardo made a point of noting that since the moon does not fall out of the sky, it must be a body to which its elements are attracted, just like earthly elements are attracted to the Earth. Space might extend infinitely beyond the known cosmos, Leonardo believed, providing ample room for other bodies like the Earth and moon.[15]

With their likening of the Earth to other celestial bodies, both Cusa and Leonardo implied a new usage for the term *world*, foreshadowing one of the prime implications of the greatest revolution in astronomy (and perhaps science): the heliocentric theory of the cosmos proposed by Nicolaus Copernicus.

ON THE REVOLUTIONS

Copernicus was born in 1473 in Torun, Poland (although he probably grew up speaking German). His father died when Nicolaus was 10. But he had the good fortune of having a prominent uncle (who would become a bishop), and so a career for Copernicus in the church was assured. After attending the University of Kraków, Copernicus was elected to be a church official in Varmia, his uncle's bishopric. But he delayed taking up his position there in order to study church law in Italy at the University of Bologna. In truth, Copernicus had been most intrigued by math and astronomy while at Kraków, but he realized he needed to go to Italy to master the cosmos at a more advanced level. In the year 1500, he visited Rome to view an eclipse of the moon. A year later he received permission to extend his studies yet further, this time to study medicine in Padua (later in life he served as his uncle's doctor). Ultimately Copernicus earned his doctoral degree in church law at the University of Ferrara in 1503.[16]

Copernicus then finally reported for his church duties in Varmia, at first serving his uncle directly at Lidzbark Warmiński for several years before joining the cathedral in Frauenburg (present-day Frombork). But all along he continued his astronomical observations and speculations. He even built a small observatory, a tower he cobbled together from 800 building stones. By 1510 or so he had formed in his mind the outlines of his new system of the cosmos, in which the sun took its proper place at the center. He prepared a manuscript describing the idea with limited distribution to a few friends (apparently requesting secrecy, for he did not even include his name on the manuscript). Decades later, Georg Rheticus (1514–1574), a young astronomer-mathematician, visited Copernicus and persuaded him to publish the full story. Rheticus became the only student that Copernicus ever taught. Together they worked on preparing the complete manuscript; Rheticus arranged for it to be printed in Germany, under the title *De Revolutionibus Orbium Coelestium* (*On the Revolutions of the Heavenly Spheres*).

It's hard to say exactly what led Copernicus to propose his radical reordering of the spheres. He never explicitly said. He was aware of ancient Greek speculation on the subject. He noted that one pre-Aristotelian philosopher (Philolaus) had suggested motion of the Earth (around a central fire) and that others had suggested a rotation on its axis. Copernicus was probably not aware of the more detailed heliocentric views of Aristarchus of Samos (d. 230 BC) or of the medieval writings of Nicole Oresme and others articulating the view that the Earth rotated on its axis.

(And in any case Oresme had not suggested Earth's revolution around the sun.) Astronomy historian Owen Gingerich has noted that there is simply no "clear-cut path" that led Copernicus to his view of the cosmos. "Copernicus's cosmology was a great adventure of the mind, a mental construction not forced by any observations and in fact contrary to the immediate senses," Gingerich wrote.[17]

It is clear, though, that part of Copernicus's motivation was simply to improve the descriptions of the heavens provided by earlier astronomers. He pointed out that Ptolemy, for instance, required all sorts of mathematical juggling to get circular motions to produce the proper predictions for the positions of the stars and planets. Copernicus praised Ptolemy as the ancient astronomer "who stands far in front of all the others on account of his wonderful care and industry"; yet modern observations did not always correspond with the movements that Ptolemy's system predicted. Copernicus explicitly set out to correct those deficiencies.

But even as he was rearranging the heavens, Copernicus was devoted to preserving much of the previous picture. His universe still comprised a set of concentric spheres, with the fixed stars attached to the outermost sphere, which enclosed the cosmos. And Copernicus insisted on retaining Aristotle's belief in the necessity of uniform circular motion. It was OK for the Earth to move, Copernicus said, but it still needed to move with the uniform circular motion that Aristotle had affirmed for all celestial bodies. Planets embedded on spheres still moved in great circles as their spheres rotated. Circular motion was required, Copernicus believed, to explain the recurrence of "irregularities" in planetary motions more harmoniously than in Ptolemy's system. "We must conclude that the motions of these bodies are ever circular or compounded of circles," he wrote. "For the irregularities themselves are subject to a definite law and recur at stated times, and this could not happen if the motions were not circular, for a circle alone can thus restore the place of a body as it was."[18]

For Copernicus, then, the problem of the heavens was to understand the geometry of the circular paths generated by the rotations of the spheres. Spheres accomplished their task through their mechanical motions, but the object was not understanding the physics of the spheres. It rather was formulating the mathematical description of the circular motions that the spheres produced. It was easier, simpler, more accurate, and more harmonious, Copernicus contended, to describe those motions in a system with the sun stationary at the center. And if the sun was stationary, the Earth must move.

CELESTIAL BALLET

Copernicus marshaled the arguments for Earthly motion much like a lawyer presenting a case in court. He began by establishing that the world—that is, the entire cosmos—is spherical, as are celestial bodies. And so too the Earth is itself a globe (although not a perfect sphere, given its mountains and valleys). Copernicus listed the ancient astronomical, geometric, and observational evidence for the roundness of the Earth, from the changing view of the heavens when moving from north to south to the slow disappearance of a ship's mast as it sails away from shore. Having demonstrated that the Earth is a sphere, Copernicus then set out to investigate what "movement follows upon its form" and "what the place of the Earth is in the universe." He warned that the opinion of numerous authorities placing the Earth in the cosmic center threatens to make any contrary proposal seem unthinkable or ridiculous. "If however we consider the thing attentively, we will see that the question has not yet been decided."[19]

When observers on the Earth witness a "celestial ballet" performed by the objects in the heavens—the planets and stars rising and setting daily (as do the sun and moon)—those motions seem "to involve the entire universe except the earth and what is around it." But Copernicus pointed out that the appearance of motion—observed change of place—can be caused by a movement of either the observed object or of the observer. And he contended that the heavens are not responsible for the celestial ballet: "If you grant that the heavens have no part in this motion but that the earth rotates from west to east, upon earnest consideration you will find that this is the actual situation concerning the apparent rising and setting of the sun, moon, stars and planets."[20]

To Copernicus this view made better sense than Aristotelian dogma did. Rotating the vastness of the entire universe once a day is much more outrageous than the Earth simply turning on its axis. It's easier to imagine the natural rotation of a spherical Earth "rather than put the whole world in a commotion—the world whose limits we do not and cannot know."[21]

Others (such as Oresme) had previously made that point. But a rotating Earth could still occupy the center of the cosmos. Copernicus went further, insisting that the apparent rotation of the sky around the Earth doesn't imply that the Earth is in the center of it all. As even the ancient Greeks were aware, the planets' motions seem irregular—some of them occasionally even reverse direction. And as changes in their brightness indicate, the planets do not maintain a constant distance from the Earth, which proved to Copernicus that the Earth could not be the center of their orbits. But such observations could be explained naturally if the Earth is just one of the planets, itself in motion around the sun.

Displacing the Earth from the center of the cosmos wrecked Aristotle's notion of the element earth's natural tendency to fall toward the cosmic center. Rather than the center of terrestrial gravity coinciding with the center of the universe, gravity could have multiple centers. Copernicus asserted that gravity is just the "natural desire" of matter to gather together and form a globe. The same "impulse" in matter on Earth should be operating on the sun, moon, and other planets, which is why they all take the shape of spheres. Since these bodies move through space, so must the Earth. And in fact, some of the oddities of their movements—when they reverse their paths in retrograde motion—are actually illusions caused by the Earth's movement. All of this becomes clear, Copernicus declared, "if only we face the facts . . . 'with both eyes open.'"[22]

From these considerations Copernicus produced his new map of the heavens. The moon, to be sure, still revolved around the Earth. But the Earth now revolved around the sun, as did the other planets. Copernicus calculated the distance from the sun to the planets (relative to the sun-Earth distance), correctly noting that the distance from planet to planet was appreciable. But those distances were dwarfed by the enormous space intervening between the planets and the sphere of the fixed stars, compared with which the Earth-sun distance was imperceptible. (It was this great distance that refuted one obvious objection to Copernicus's system: that the positions of the stars should change because of parallax as the Earth orbited the sun. Copernicus replied that with the stars at such a great distance, such a parallax effect would be too small to detect.)

Whether the universe was truly infinite or not, he said, was a question he preferred to leave to the philosophers of nature. Copernicus himself was no philosopher but rather a mathematician and astronomer. In those days, in fact, astronomer and mathematician were almost interchangeable terms. Astronomy was in essence a form of mathematics, a method for saving the phenomena, not explaining the physical processes underlying them.

In this respect there is one aspect of Copernicus's work that is especially pertinent to today's multiverse controversy. This issue arose in the introduction attached to Copernicus's *De Revolutionibus* by Andreas Osiander, a German theologian who helped oversee its printing. Osiander took it upon himself to declare in the (unsigned) introduction that the work presented "hypotheses" about the heavens, meaning that it merely offered a mathematical method for predicting the motions of celestial objects. It was not to be taken seriously as a physical description of the cosmos. "It is the job of the astronomer to use painstaking and skilled observation in gathering together the history of the celestial movements, and then . . . think up or construct whatever causes or hypotheses he pleases such that, by the assumption of these causes, those same movements can be calculated from the principles of geometry," Osiander wrote. The astronomer cannot "by any line of reasoning reach the true causes of these movements." For "it is not necessary that these hypotheses should be true, or even probable; but it is enough if they provide a calculus which fits the observations."[23]

Anybody who actually read the rest of the book should have realized that Copernicus did not write that introduction—he clearly believed in the physical truth of his sun-centered planetary system. In fact, legend has it, Copernicus was literally mortified when the printed copy of his book was delivered to him while on his deathbed. When he saw Osiander's introduction, some historians suspect, it induced a stroke that finished him off.

Yet while Osiander's introduction misrepresented Copernicus, it did raise an important issue about science that influences debates about the universe today. Mathematics is necessary for describing reality accurately—that is, for representing it in a way that reproduces and predicts observed phenomena. But does that mean that the mathematics coincides with something physically real? Or is the math just a useful guide, a map of nature that gets you where you want to go but not actually a true depiction of reality? In technical philosophical terms, the question is whether the math is ontic or epistemic—is it real, or is it informational? Roughly, it's the distinction between real music—waves vibrating in the air—and the symbols for notes on a sheet of paper that provide the information you need to make the music. It's similar to the question from chapter 2 about whether Eudoxus meant his spheres merely as mathematical devices for calculating the paths of heavenly bodies (epistemic) or real physical entities carrying those bodies around (ontic).

Exactly this question is debated today with respect to the mathematics of quantum theory. Does that theory's key mathematical object, the wave function, correspond to something that really exists in the physical world (ontic), or is it merely a convenient calculational tool, describing knowledge about nature's behavior (epistemic)? The answer to that question is at the heart of the case for the Many Worlds Interpretation of quantum mechanics, one of the most intriguing modern versions of a theory of a multiverse, positing a quantum plurality of worlds. (We'll visit that multiverse in chapter 11.)

Unlike Cusa before him (and Giordano Bruno to follow), Copernicus did not directly address the medieval question of the plurality of worlds. Yet Copernicus plays a major role in the plurality-of-worlds debate, in two essential ways. First, he redefined the concept of the universe. No longer a set of spheres with the Earth at the center, the universe was now a solar system, a set of planets orbiting the sun, surrounded at a great distance by a sphere of stars. Second, the idea of "world" no longer necessarily coincided precisely with the notion of the universe. "World" could now refer to any of many celestial bodies, of which the Earth was one. With the establishment of the Copernican solar system, at least a few other "worlds" now automatically existed. So with Copernicus, the issue forks: more "worlds" do exist, but whether multiple universes exist remains unknown. Whether other solar system "universes" populate the heavens—that is, whether the stars are other suns, with planetary retinues of their own—was soon to be investigated by Copernicus's followers.

Wandering in Immensity

I believe that there are infinite worlds, each of them, however,
being finite, just as [that] whose middle point of the planets
is the center of the Sun. As with the Earth, the Sun does not rest,
since it revolves most swiftly in its place around its own axis. . . .
The stars are also moving like the Sun, but they are not carried
around it by its force, because each of them is a Sun in a world
that is not smaller than our own world of the planets.

—EDMUND BRUCE, LETTER TO KEPLER, 1603

GIORDANO BRUNO should have been a character from Harry Potter's world of wizardry. He believed in magic.

In the middle of the sixteenth century, magic was a big deal for many European philosophers. Some of them believed that the ancient Egyptians possessed magical powers and knowledge from which the secrets of nature could be discovered. As a young man, Bruno fell under the spell of the "natural magicians" and developed his own views about the implications of magic for the universe. For Bruno, heavenly bodies were living things, like animals, moving through space. And he believed in an infinity of such animals, populating an infinite space.

Bruno is naturally a controversial figure in the history of science, as he was in his own time. He was burnt at the stake for heresy (although perhaps not for his Copernican cosmology or belief in an infinite universe). The eminent twentieth-century science historian Alexandre Koyré considered Bruno to be "not a very good philosopher" and "a very poor scientist."[1] Nevertheless Bruno advocated a conception of the cosmos that eventually became the scientific consensus, at least in its broad outlines.

He was not the first post-Copernican scholar, though, to promote the idea of an infinite universe. In 1576, Thomas Digges in England translated major portions of Copernicus's *De Revolutionibus*. Digges noted that Copernicus himself had refused to speculate on the possibility of infinite space. Digges was not so reluctant.

ALTITUDE WITHOUT END

Born about 1546—three years after Copernicus published *De Revolutionibus*—in the county of Kent in England, Digges was the son of a well-known mathematician, Leonard Digges.[2] It's doubtful that Thomas attended any university. He was home-schooled in math until his father died when Thomas was a teen. Thereafter the mathematician-astronomer John Dee (1527–1609) assumed the role of Digges's tutor-mentor. Dee, who knew the work of Copernicus, was famous for his service as astrologer to Queen Mary and later as an adviser to Queen Elizabeth.[3] Dee was also friends with John Field, who in 1557 published the first ephemeris (a book giving the positions of the planets) in England using the Copernican system.[4] Digges was the first, though, to openly defend and articulate the wisdom of Copernicus to the English public.

Very much like a modern scientific thinker, Digges performed experiments related to navigation and ballistics in addition to his astronomical research, at which he excelled. "Digges was one of the ablest and most accurate astronomical observers of his time," scholars Francis Johnson and Sanford Larkey declared.[5] Digges was no doubt introduced to Copernicus by Dee, who had written a foreword for Field's 1557 ephemeris.

Thomas's father, Leonard, had published a successful almanac called *A Prognostication Everlasting*, which included a map of the heavens based on Earth-centered Ptolemaic astronomy. For a new printing of the almanac in 1576, Thomas realized that it needed an addendum depicting and describing the Copernican system, "so that Englishmen might not be deprived of so noble a theory."[6] In an appendix to the almanac, Digges presented a translation of key chapters from *De Revolutionibus,* with a few comments and elaborations of his own.

At the time, few knew that Osiander's foreword to *De Revolutionibus* had not been written by Copernicus himself. But Digges, having read the book insightfully, realized that Copernicus did in fact intend his model to be taken as physically correct. Copernicus "meant not as some have fondly excused him to deliver these grounds of the Earth's mobility only as mathematical principles, feigned and not as philosophical truly averred," Digges wrote.[7]

Copernicus had indeed rearranged the world. But it was still a finite world of nested crystalline spheres. He didn't wonder or worry whether other sets of spheres might exist beyond. For Copernicus, the universe (or world) was the solar system, encased by the very distant sphere of the fixed stars. Digges, on the other hand, saw that Copernicus could or should have gone further. Digges decided that the realm beyond the planets would be a perfect space for an abode for the angels, illuminated by stars too plentiful to count. The sphere of the stars did not limit the world, which extended ever upward. "We may easily consider what little portion of god's frame, our elementary corruptible world is," Digges wrote, "but never sufficiently be able to admire the immensity of the rest. Especially of that fixed orb garnished with lights innumerable and reaching up in spherical altitude without end."[8]

In Copernicus's cartoon of the cosmos, the outer sphere of fixed stars was depicted as a boundary enclosing the world. Digges, to the contrary, produced an image with stars scattered outside that boundary, stars that could extend to infinity—spherical altitude without end. Most scholars interpret Digges to be the first Copernican to declare the universe to be infinite in extent—the

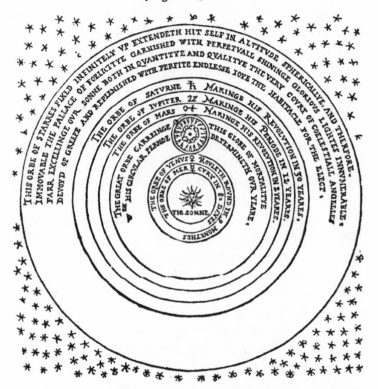

Thomas Digges's Solar System. Thomas Digges, an early advocate for the Copernican view of a sun-centered planetary system, illustrated the order of the planets much as Copernicus did. But whereas Copernicus assigned the stars to an outermost sphere at a fixed distance from the sun, Digges depicted stars outside the outer sphere, indicating his belief in an immense space beyond the solar system.

question Copernicus had left to the natural philosophers. "Digges . . . definitely maintains in his work that the universe is infinite, that the stars are numberless, and that they are located at varying distances from the center, the sun, and extend throughout infinite space," Johnson and Larkey wrote in 1934 in an important paper that reprinted Digges's translation.[9] Before this paper appeared, most historical accounts credited the original idea of an infinite universe to Giordano Bruno.

AN INFINITY OF WORLDS

Whether the universe is infinite had, of course, long been a popular philosophical question, linked closely with the debates about a plurality of worlds. Renaissance writers were well aware of the controversies surrounding those issues. In the same year that Digges published his Copernican appendix, for instance, French theologian Lambert Daneau, in a treatise called *The Wonderful Workmanship of the World,* explored the plurality-of-worlds question in dialog form. In one passage, a student poses a question to the master, asking about those who defend the idea "that there be many worlds." The master replies that followers of Democritus and Epicurus contend that just as we occupy the Earth, other beings inhabit other worlds like ours, with their own moon, sun, and heavens. Some philosophers even suppose that there are an infinite number of such worlds, the master said. And so, asks the student, "Are there indeed many worlds?" And the master replies: "Fie upon this infinite or multitude of worlds. There is one and no more"—foreshadowing the attitude of many scientists today.[10]

Giordano Bruno believed otherwise. Born in Italy in 1548, Bruno became a Dominican friar at Naples, where Thomas Aquinas had taught three centuries earlier. While in Naples, Bruno developed prodigious skill at memorization, which apparently was part of being a magician. Bruno's inclination to unconventional thinking did not go unnoticed, and he faced accusations of heresy while in his 20s. Eventually he decided to quit the Dominicans and leave Italy, thereafter moving from country to country as he developed and promoted his philosophy.[11]

He had some success in France, winning support from the king (who seemed impressed by Bruno's magical memory skills). With a recommendation from the French king, Bruno moved on to England, where he lived and wrote from 1583 to 1585. While in England he composed defenses of the Copernican system. Like Copernicus, Bruno viewed heavenly bodies as self-gravitating centers, each in a way a world unto itself. All were conceived by Bruno to be magical planetary "animals." In one of his most important works, the 1584 treatise *De l'infinito universo et mondi* (*On the Infinite Universe and Worlds*), Bruno insisted that these worlds are bodies, heterogeneous globes, in which earth "is no more heavy than the other elements; all the parts and particles are moved and change place . . . as do the blood, humours, spirits, and insensible particles which perpetually flow in and out of us and in the other lesser animals."[12]

Bruno took this idea to the extreme—his universe contained an infinity of such bodies, and hence was itself infinite, a direct manifestation of the infinity of God. Influenced by Nicholas of Cusa's similar writings, Bruno even considered the likelihood that these worlds possessed inhabitants. His view of an infinite cosmos was deeply religious, stemming from his belief in God's own infinite presence. "Thus is the excellence of God magnified and the greatness of his kingdom made manifest; he is glorified not in one, but in countless suns; not in a single earth, a single world, but in a thousand thousand, I say in an infinity of worlds," Bruno asserted in *De l'infinito*.[13]

Bruno's beliefs stemmed not just from Copernicus and Cusa, but also from a reaction to the philosophy of the ancient Stoics. Like the atomists before them, the Stoics believed in an infinite universe. But the Stoics' universe was mostly empty. Stoics believed that the "world"—Aristotle's set of spheres—was spherical and finite, but that a vast infinite expanse of nothingness surrounded it. That was not good enough for Bruno. His infinite space needed to be full of stars. "Infinite perfection is far better presented in innumerable individuals than in those which are numbered and finite," Bruno declared.[14]

He favored something closer to the views of the atomists, which he probably learned from reading Lucretius. So for Bruno, "worlds infinite in number are distributed through a universe of infinite extent."[15] And the individual heavenly bodies, moving on their own (because they are alive), had no need for spheres. In this respect Bruno agreed with Tycho Brahe, who had decided on his own in the 1570s that Aristotle's spheres got in the way of working out heavenly motions rather than help explain them.

THE IMMENSITY OF EMBOSOMING SPACE

While Thomas Digges preceded Bruno by a few years in proclaiming the universe to be infinite, his impact on the rest of the scientific world was not as great. Digges's writings were widely known in England but not elsewhere (primarily because he wrote in Elizabethan English rather than Latin). It didn't help that his major exposition of Copernican thought was an appendix to a popular almanac rather than in some sort of explicitly scientific treatise. Koyré, writing in the mid-twentieth century, didn't even think Digges had conceived of cosmic infinity in quite the modern way. It seemed to Koyré that Digges was contemplating more of a theological infinity, where the angels would have plenty of space. But Koyré acknowledged that "Digges was the first Copernican to replace his master's conception, that of a closed world, by that of an open one."[16]

Other scholars have argued that not only did Digges precede Bruno with the notion of an infinite universe, but also that Bruno may have gotten the idea from Digges. When Bruno visited England from 1583 to 1585, Digges's work was widely known there. And Bruno had not written anything about the infinity of the universe before then. "It is entirely possible that Digges's brief treatise on the Copernican system first suggested to Bruno's mind the thought of using the new heliocentric theory as a physical proof of his highly speculative notions concerning the infinity of the material world," Francis Johnson wrote.[17] Historian Stillman Drake also noted that "Bruno's first, and perhaps his only, acquaintance with the doctrine of Copernicus came through" Digges's work.[18]

Koyré nevertheless insisted that it was Bruno who, "for the first time, presented to us the sketch, or the outline, of the cosmology that became dominant in the last two centuries."[19] And in fact, Bruno did take a decisive step beyond the infinity proposed by Digges: proclaiming the existence of a multiverse.

Bruno did not use that term, or course. He spoke of many worlds. While for Digges there existed an infinity of stars, they merely occupied the outer reaches of the single world, the sun-centered system of planets (including the Earth). That world—planetary system plus distant stars—was the universe, as conceived at that time. Before Bruno, all conceptions of the universe-world, whether Aristotelian or Copernican, viewed the system containing the sun and Earth as special, one of a kind. "All these cosmic representations had one point in common: a profound heterogeneity between the unique planetary region and the region . . . of fixed stars," historian Miguel Granada has pointed out.[20]

Bruno, on the other hand, drew a distinction between the universe (*universo*) and world (*mondo*). "World" still refers to the perceived universe (what the Greeks called the *kosmos*). "Our" world is the "universe" that we see—the planets and stars, "occupying our space, bounded by the vault of heaven." But beyond exists an infinite space that includes countless other such worlds, additional "systems of heavenly bodies, each system occupying its own space."[21] Each world possesses its own central sun and planets (or "earths"); the various worlds are separated by vast distances, extending to infinity.[22]

That infinity, wrote Bruno, comprises the one whole greater universe. It "is one, the heaven, the immensity of embosoming space, the universal envelope, the ethereal region through which the whole hath course and motion."[23] Just as we can perceive many suns, Bruno says, we can infer the existence of an infinity of other such systems. There was simply no reason to believe otherwise.

"The fixed stars, those magnificent flaming bodies, are . . . in-numerable worlds, not greatly unlike the world in which we live," he wrote. "Let not this contemplation dispirit man, as if he thought himself abandoned by God; for in extending and enlarging the universe he is himself elevated beyond measure, and his intelligence is no longer deprived of breathing space beneath a sky, meager, narrow, and ill-contrived in its proportions."[24]

So Bruno really was an advocate for a multiverse: an infinity of worlds, each world composed of many systems of stars accompanied by "earths." He simply then redefined the "universe" as the totality of all such worlds: the universe became the entirety of existence, and it was infinite, and there could be only one. Bruno's terminology exemplifies the recurring theme of the multiverse debate—the recognition that the world, or universe (as conceived at the moment), is just one of many such worlds, components of a grander universe. Once the reality of hypothesized multiple worlds becomes established, the term "universe" assumes a new meaning.

An important part of Bruno's reasoning involved distrust of the senses. In reading Cusa, Bruno perceived an important shift "from sensual to intellectual cognition."[25] When it comes to cosmic questions, Bruno believed, the senses cannot be trusted; they are subject to confusion and illusion. If the universe is indeed infinite, then the senses are a poor guide, as the infinite is not accessible to them. It must be analyzed by the intellect.

"No corporeal sense can perceive the infinite," says Philotheo, the character representing Bruno in one of his dialogs. "None of our senses could be expected to furnish this conclusion; for the infinite cannot be the object of sense-perception." Bruno assents that there is need for some evidence from the senses, at least in regard to "sensible objects." But "even there it is not above all suspicion unless it cometh before the court aided by good judgment."[26] It's the job of the intellect to weigh the evidence of the senses along with other considerations.

In response to Philotheo, the dialog character Elpino asks what the senses are good for, then. Philotheo replies: "Solely to stimulate our reason, to accuse, to indicate, to testify in part; not to testify completely.... Truth is in but very small degree derived from the senses as from a frail origin, and doth by no means reside in the senses." And if not in the senses, Elpino asks, where? Philotheo replies: "In the sensible object as in a mirror. In reason, by process of argument and discussion. In the intellect, either through origin or by conclusion. In the mind, in its proper and vital form." Thus persuaded, Elpino (in a later dialog) sums up the lesson: "There are then innumerable suns, and an infinite number of earths revolve around those suns, just as the seven we can observe revolve around this sun which is close to us." And Philotheo responds: "So it is."[27]

WANDERING IN IMMENSITY

Bruno, unlike Digges, was certainly widely read by the scientists of Europe. But Bruno's views were not initially met with widespread approval. In fact, one of the most prominent of his readers, Johannes Kepler, disagreed vigorously. Kepler was appalled by the vastness of an infinite cosmos—it gave him nightmares, at least figuratively.

"The mere thought brings with it I know not what of secret horror," Kepler wrote. "One finds oneself wandering in this immensity, whose boundaries, whose center, and thus any defined place at all, are all denied to exist."[28]

Kepler did not disagree, of course, with Bruno's heliocentrism, having learned the Copernican view of the universe as a student after an unpromising youth. Kepler was born into poverty. As a child he worked in his father's tavern until bankruptcy forced him to quit school and do field labor. On the bright side, his impoverished situation qualified him for a scholarship at a theology school while a teenager. From there he advanced to a college where he did so well he was accepted for graduate school at the University of Tübingen.

Among the professors at Tübingen was Michael Mästlin (1550–1630), an early adopter of Copernicanism. Under Mästlin, Kepler studied astronomy but retained an interest in theology and contemplated a church career. Fortunately for the history of astronomy, his first job offer came from a school in need of an astronomy professor. So Kepler moved to Graz in 1594, a first step toward becoming the preeminent astronomer of the early seventeenth century.[29]

From the beginning, he imagined that God had a grand geometric plan in mind when creating the solar system. So Kepler constructed a model of planetary positions based on the five regular polyhedra of solid geometry. His writings on the topic impressed the astronomical community—in particular, Tycho Brahe, the Danish astronomer who had elevated his science dramatically by performing the most accurate measurements ever made of the heavens. In 1600 Tycho, then working at Prague, invited Kepler to be his assistant. Tycho died soon thereafter. So Kepler took over Tycho's job as imperial mathematician for Holy Roman Emperor Rudolf II and took possession of Tycho's meticulous observations.

While his personal life was a mess—his family plagued by illnesses and deaths—Kepler still managed to modernize the Copernican solar system by establishing that the planets' orbits are ellipses, not some combination of circles. Later he took a position in Linz, Austria, and published further important astronomical treatises plus an important set of numerical tables for use in making astronomical calculations.

Later in life (in fact, the book was not published until four years after his death in 1630), Kepler wrote the *Somnium* (*The Dream*), a science-fictionish account of life on the moon. So Kepler, like Bruno, envisioned intelligent beings on other worlds. But even though Kepler held Bruno in high regard, an infinity of others beyond the known worlds went too far. "As far as I am concerned," Kepler wrote, "it is certainly not good to roam in that infinity."[30]

Kepler insisted on the view that astronomers dealt with appearances—phenomena that could be perceived. Kepler believed that hypotheses about celestial motions must not transcend the phenomena "by positing the existence of things that are either incompatible with them, or, even worse, of things that do not and cannot 'appear.'"[31]

According to Koyré, Kepler's beef with Bruno was the assumption of a uniform cosmos, an infinite space everywhere similar, with worlds scattered about without structure. Bruno held that God has no need to distinguish any one part of the universe from another. For Kepler, "an irregular, irrational scattering of fixed stars in space is unthinkable; finite or infinite, the world must embody a geometrical pattern."[32] But an infinite space implies lack of such structure, in Kepler's view.

Kepler did not insist that no stars existed beyond those known, or that those that could be seen were all at the same distance from the sun. "It is rather uncertain," he wrote, "since some are small, some are big. However, it is not unlikely that the small ones seem so because they are withdrawn high in the ether, and the big ones are nearer to us."[33]

But even if unseen stars exist, Kepler contended, they do not imply infinity. Many stars might be too small or too distant to be visible without space extending infinitely. And even if stars exist vastly far away, they do not extend their presence "downward" into the realm near the Earth. Since we see only half of the Milky Way at one time, it surrounds us at a distance. The sphere of the fixed stars is limited, held away from us by both "the stellar orb" and the "circle of the Milky Way," Kepler contended. It is this evacuated neighborhood, where the sun and planets live, that demonstrates the nonuniformity (and hence noninfinity) of the world. We live in a special place, privileged above the rest of the universe. (He referred to our solar system as "the very bosom of the world."[34]) It is meaningless for science to concern itself with anything other than our peculiar realm of the cosmos and the stars that we see. "It is certain that inwards, toward the sun and planets, is a finite world, in some way excavated," Kepler asserted. "What remains is superfluous metaphysics."[35] For Kepler, the concept of "universe" was restricted to the "observable universe."[36]

In this respect Kepler's argument is very much like those of today's multiverse deniers. Even if stellar worlds exist beyond our sight, as Bruno contended, those worlds have no relevance to science. "If they are not seen," Kepler declared, "they for this reason are not pertinent to astronomy."[37] As Koyré summarized Kepler's views, it's fine to believe, if you like, that countless stars—maybe most of the stars—are too far away to see. "But it will be a purely gratuitous assumption, not based on experience, that is, on sight. These invisible stars are not an object of astronomy and their existence cannot in any way be demonstrated."[38] If you replace "stars" with "universes" you would be making the exact same argument that multiverse opponents do today. It's one of the reasons why the multiverse argument is not only about cosmology but is also about the nature of science itself.

Kepler clearly did not fully appreciate Bruno's point about the need for intellect to complement the senses. Modern scientists who argue that the multiverse is nonsense because other universes cannot be observed are also ignoring Bruno's insight. This is an important point for doing science, in particular cosmology. The right way to do science depends on the nature of reality. If the universe truly is infinite, science cannot be successfully done relying on the senses alone. If a multiverse truly exists, then scientists must reexamine their methods for studying reality.

Kepler expressed his opposition to Bruno's infinity a few years after Bruno was burnt at the stake in 1600 for heresy. Historians have long mostly agreed that Bruno's belief in an infinity of worlds played a small role, if any, in the case against him (although some recent scholarship questions that conclusion).[39] In any event,

Bruno's religious views were undoubtedly too unconventional for the church, and his affinity for magical practices also worked against him. His execution probably did, however, negatively influence the willingness of other Copernicans to speak out about the motion of the Earth. Galileo was the exception, but even he agreed to recant his opinion in exchange for house arrest rather than a penalty more severe.

As for infinity, Galileo did not seem to be very interested in that debate. He did mention that the stars are suns, but did not contend that they are scattered through an infinite space. In his famous *Dialogue Concerning the Two Chief World Systems,* his character Salviati mentions that nobody "has so far proved whether the universe is finite and has a shape, or whether it is infinite and unbounded."[40] In a letter to his friend Fortunio Liceti, Galileo said that none of the arguments pro and con about the infinity of the universe led, in his mind, to a necessary conclusion. And his imagination was of no use in determining the answer because "I do not know how I could imagine it either limited, or unlimited and infinite." But he did suggest that infinity is more incomprehensible than finiteness. Still, Galileo concluded that whether the universe is infinite or not "is one of those questions that happen to be inexplicable by human reasonings."[41]

FRAGILE REED OF WILD IMAGINATION

Infinity's reality (or not) had always been an important aspect of the question of the plurality of worlds. It had been a key point of disagreement between Aristotle and the atomists. But the other major tenet of atomist thought—the existence of atoms—had not played much of a role in the medieval debates among the scholastics. Natural philosophers in those days knew of atomism only through Aristotle's criticism of the atomists. But by the end of the medieval period, atomism had begun to return to the scientific scene, and it would once again become a prominent part of the debate over infinity and multiple worlds.

While the atoms of the ancients built the Greek *kosmos*, modern atoms adapted to the new notion of "world" as a sun accompanied by planets, as Steven Dick has pointed out. "As the Aristotelian tradition gradually gave rise to a concept of 'world' as a center of attraction that could be identified with each of the celestial bodies, so atomism would assimilate that idea and, through its preoccupation with cosmogony, introduce the concept of a solar system as a 'world.'"[42]

More complete knowledge of ancient atomism became possible after 1417, when Poggio Bracciolini, a papal secretary, found a manuscript copy of Lucretius's *De Rerum Natura* in a German library.[43] Lucretius's work made no immediate impact on science—though it was admired as a nice poem. But toward the end of the sixteenth century, atomism began its revival, with the new awareness of Lucretius one of the driving forces. Further support came from the publication of Diogenes Laertius's *Lives of Eminent Philosophers* in 1533, which provided more detail about Epicurus. And as atomism rose again, Dick says, "with it came a rejuvenation of interest in the concept of an infinite number of worlds."[44]

In 1563 a new version of Lucretius's poem was published in Paris, where it caught the eye of the famous essayist Michel de Montaigne. In his *Essais* (1580s) Montaigne indicated sympathy to the universe of the atomists, with its uncountably many worlds. "Your reason is never more plausible and on more solid ground than when it convinces you of the plurality of worlds," Montaigne wrote. "The most famous minds of times past have believed this."[45]

Though not a scientist himself, Montaigne raised a key point about multiple worlds that echoes in today's debates: whether all those many worlds are alike. "Now if there are many worlds, as Democritus, Epicurus, and almost all philosophy has thought, how do we know whether the principles and rules of this one apply similarly to the others?" Montaigne asked. "Perchance they have a different appearance and different laws."[46] In today's multiverse debates, the notion of many worlds with different laws, or at least different physical properties, is the key issue regarding whether the multiverse can explain the properties of the universe we live in. If all the multiple universes are identical, our universe can no longer be explained as the Goldilocks version, where all its features are just right for life.

In the early seventeenth century, the atomist cause was taken up more rigorously by Pierre Gassendi (1592–1655). Gassendi, born in Champtercier, France, was an intellectually gifted youth who studied philosophy and theology in his teen years and by age 21 found himself teaching rhetoric at the College of Digne. Soon thereafter he earned his theology doctorate and was ordained a priest in 1616. He then returned to Aix-en-Provence (where he had studied as a youngster) to become a professor of philosophy. Among his many interests was astronomy. After a few years at Aix-en-Provence, Gassendi moved back to Digne. He often visited Paris, where he interacted with many of the leading scholars of the day.[47]

While at Digne, Gassendi began a thorough study of Epicurus, planning to eventually write a treatise on Epicurean philosophy. It became a long-term project, as Gassendi was also busy making astronomical observations and composing various treatises on other topics, including alchemy.

Influenced by his study of Epicurus, Gassendi became a vigorous advocate of atomism. He even imagined that one day microscopes would be powerful enough to view atoms directly (and three centuries later, he was right). But Epicurean philosophy was not exactly religiously correct, and Gassendi, well aware of the fates of Bruno and Galileo, was careful to frame his atomism so as to keep it compatible with Catholic dogma.

Unlike the Greeks, who envisioned an infinite number of atoms creating an infinite number of worlds, Gassendi allowed that God had created a finite number of atoms with which to make the visible world. Gassendi did view our visible world as sitting in a surrounding infinite space, essentially adopting the view of the Stoics. But he amended the Stoics' view by allowing vacuum, where the

Stoics' space was completely filled by some sort of plenum. Gassendi mixed and merged the ancient atomist philosophy with that of the Stoics. In doing so he defended and propagated the notion that empty three-dimensional space was a meaningful concept—no body need occupy such space for it to exist, as Aristotle had reasoned.

Gassendi concluded from this line of thought that an immense infinite space (*spatia immensa*) must have already been in place before God created the bodies that occupied it. This preexisting space would be itself motionless and impose no resistance to the motion of bodies within it. It is, in fact, very much like the absolute space that Isaac Newton was to describe a few decades later. Gassendi added that time as well was independent of the bodies in the universe. "Even if there were no bodies, there would still remain both an unchanging place and an evolving time." Gassendi, wrote Edward Grant, "constructed the spatial framework of Newtonian physics and cosmology."[48]

In his *Syntagma Philosophicum* (published after his death), Gassendi took up the question of plurality of worlds, stressing that "world" and "universe" were no longer synonymous, as they had been pre-Copernicus (or at least pre-Cusa). As Gassendi framed it, the question to ask is whether the world *is* the universe—that is, whether there is one world in the universe or many. Gassendi's discussion showed that in the post-Copernican era the nature of the question was changing, and sometimes old views were mingled with new ones in attempts to answer it.[49] Gassendi noted one important distinction: between multiple worlds and an infinity of worlds. And he also discussed the old issue of a plurality of worlds in the context of one succeeding another in time, versus a plurality of coexisting worlds.

Gassendi argued against the infinite separated worlds envisioned by Epicurus, maintaining that religious faith was consistent with only one world. God could make many worlds, perhaps, but he did not have to do everything that he could possibly do. And there is no reason to believe that he did make other worlds, although Gassendi acknowledged that there was no way to prove the contrary. In other words, Gassendi believed the question of plurality of worlds could not be answered on scientific grounds.[50]

In presenting Gassendi's views in 1654, the English philosopher Walter Charleton (1619–1707) strongly condemned the idea of multiple worlds. He allowed that God could make many worlds. But to argue that because it is possible, such worlds actually do exist, is "a conclusion repugnant to all the inducements of persuasion." Advocates of one world, Charleton wrote, "ground our opinion upon that stable criterion, our *sense,* and asserting the singularity of the world, discourse of what our sight apprehends." Multiple-universe advocates, though, base their view "upon the fragil reed of wild Imagination, and affirming a plurality discourse of what neither the information of their sense, nor solid reason, nor judicious authority, hath learned them enough to warrant even conjecture."[51] Which, although more poetical, sounds very much like what multiverse opponents say today.

A more moderate view was expressed by another Englishman, the philosopher Henry More (1614–1687). More was intrigued by the link between a plurality of worlds and the atoms of the ancients. He was a Platonist but believed it was possible to wrangle Democritus and Plato into a consistent system containing the atomists' infinity of worlds. Such an infinity of worlds, More conceded, was "a thing monstrous if assented to, and to be startled at." But he insisted that reliance on reason—not just the senses—would make such an idea plausible. (An echo of Bruno.) More was careful to insist that he did not advocate the pleasure-seeking aspects of Epicurean philosophy ("I detest the sect of Epicurus for their manners vile"), yet he had sworn a "more faithful friendship with Truth than with myself" and so could not reject the physics of Epicurus that was in accord with reality. More viewed each of the fixed stars as a sun, with its own planets, and in that sense advocated multiple worlds. "I will not say our world is infinite," More wrote, "but that infinitie of worlds there be."[52]

Later in life, More seemed to back off, rejecting his earlier belief in an infinity of worlds. Instead, he adopted the Stoics' idea of our world surrounded by an infinite space. More's later views were intermingled with his developing theology and his conception of God and were specifically in opposition to one of the great thought leaders of the seventeenth century, René Descartes. Descartes's philosophical system did not survive the later advances in physical science. But it did establish an intellectual basis for the widespread popular belief in a plurality of worlds.

Planets and People

When the heavens were a little blue arch, stuck with stars;
methought the universe was too strait and close, I was
almost stifled for want of air; but now it is enlarged in
height and breadth, and a thousand and a thousand
vortexes taken in; I begin to breathe with more
freedom, and think the universe to be incomparably
more magnificent than it was before.

—BERNARD LE BOVIER DE FONTENELLE, *CONVERSATIONS
ON THE PLURALITY OF WORLDS,* 1686

N *A STUDY* in *Scarlet,* Sherlock Holmes reveals to his sidekick Dr. Watson that cosmological mysteries are not among his specialties. Holmes, it seems, had no idea that the Earth revolves around the sun, rather than vice versa. "He was ignorant," Watson lamented, "of the Copernican Theory and of the composition of the Solar System. That any civilized human being in this nineteenth century should not be aware that the earth travelled round the sun appeared to be to me such an extraordinary fact that I could hardly realize it." After informing Holmes of this astronomical fact, Watson's shock was compounded when the great detective replied that he would now attempt to forget it. His brain needed the space for more important things. "You say that we go round the sun," Holmes said. "If we went round the moon it would not make a pennyworth of difference to me or to my work."[1]

Had Watson lived in the decades after Copernicus published his theory, in 1543, he wouldn't have been surprised to find people who didn't know that the Earth orbited the sun. It took well over a century for Copernicus's rewrite of the universe to penetrate even the scientific psyche. But by Christiaan Huygens's time, in the late seventeenth century, most Europeans knew about the idea (even if not everyone believed it). And no doubt many had the same attitude as Holmes—it didn't matter a whit to them in ordinary life. Nevertheless, awareness of the Copernican solar system, and Earth's new status as just one of multiple worlds, incited enthusiasm among scientists and nonscientists alike thanks to an important implication: if there are other worlds, perhaps there are other people.

Nobody exploited the fascination with aliens better than the French science popularizer Bernard le Bovier de Fontenelle. He incited alien mania by composing one of the most best-selling science books in history: *Conversations on the Plurality of Worlds*. In it a philosopher-narrator recounts a series of after-dinner dialogs with the Marquise, a noblewoman of little education but considerable intellect. Over five successive moonlit evenings, the philosopher describes for her the new scientific understanding of the cosmos that had been revealed by telescopes over the preceding decades.[2]

Being French, Fontenelle's view of the new natural philosophy had been shaped most prominently by the philosophical system of René Descartes. In fact, many of the seventeenth century's great thinkers had been deeply influenced by Descartes's writings, which were rich in ideas even if often wrong in the details. One of those rich ideas envisioned a universe that did, indeed, contain an infinity of worlds.

THE NUMBER OF THE HEAVENS

Descartes was born in 1596 at La Haye in Touraine province, France. (Do not look for La Haye on a map. It has long since changed its name—to Descartes.) His mother died when René was a one-year-old; he was raised by relatives and by age ten or so was sent off to a Jesuit college. After eight years of high-quality Aristotelian education, René took a break from school and spent some time in Paris until returning to academic life at the University of Poitiers, where he studied law. (His father and others in his family had been lawyers.) But for some reason he declined to pursue a legal career and decided instead to enlist in a Dutch military school when he was 21. Two years later he joined the Duke of Bavaria's army.[3]

In the winter of 1619–1620, while en route back to the army (after attending the coronation of the Holy Roman emperor), Descartes was trapped by bad weather. He retreated from the cold into a small room where he began to contemplate the nature of knowledge and reality. Apparently one night a series of dreams revealed to him that his destiny was to create a new system of philosophy, providing a foundation for all knowledge. He spent a few more years gathering experience of the world by visiting various locales around Europe, while also reading the works of several contemporary philosophers, including Giordano Bruno.

After some time in Paris, where Descartes worked on mathematical issues and began compiling his rules for how to apply math to philosophy, he eventually moved back to Holland. There he composed his major philosophical works, with the design of deriving all of knowledge from "first principles." His dreams in Bavaria had persuaded him of the need to build a new philosophy of reality from scratch, relying on fundamental tenets from which deductions about reality could be made. Indeed, Descartes believed he had found a system of philosophy that would guarantee the truth about the nature of the world. He wrote to his friend the French mathematician Marin Mersenne (1588–1648) that he planned a treatise that would explain "all the phenomena of nature, that is to say, the whole of physics."[4]

Descartes began by composing a treatise called *The World*, in which he incorporated the Copernican view of the universe. For Descartes, the universe was a machine. Matter existed in the form of corpuscles, which interacted according to laws of motion (established by God); the world operated as a mechanism in which particles of various shapes and sizes produced the structures and actions observed in nature. Matter did not need to be infused with Aristotle's forms to take on specific properties. Change was merely the result of matter in motion, motion governed by God's laws.

When he was done with this work, Descartes learned of Galileo's conviction for promoting Copernicanism and therefore withheld *The World* from publication. He subsequently published some of his ideas anonymously. Later his major works, which contained many of the ideas he had earlier withheld, began to appear: *Discourse on Method* in 1637, *Meditations on First Philosophy* in 1641, and in 1644 *The Principles of Philosophy*, which gave a more complete presentation of his system of physics. (*The World* was published in 1664, long after his death in 1650.)

In his later writings, Descartes clarified that his corpuscles were not the atoms of the ancient Greeks—he felt that matter could be divided infinitely without reaching any uncuttable particle. (At least, any particle could be further divided "in thought" if not in actuality, and God in any case could always further divide any particle, no matter how small.) And where Democritus and his followers believed that atoms moved in a void, Descartes insisted that there could be no void; all of space was filled with fluidlike matter, a plenum. In other words, all of space is corporeal. In fact corporeality—or matter—is just the extension of space in three dimensions, Descartes contended. "The idea of extension which we conceive in any space whatever," he wrote, "is plainly identical with the idea of corporeal substance." And it is equally obvious (to Descartes) that three-dimensional space cannot be bounded. "This world, or the whole (*universitas*) of corporeal substance, is extended without limit," he wrote.[5] If we try to imagine such limits, we can always then imagine something beyond, and thus believe such extended space to be real as well. And so for Descartes the universe is at least indefinitely extended, if not infinitely extended (like Cusa, he reserved the truly infinite for God).[6]

Within this plenum, Descartes believed, matter observed certain laws of motion. He was on the right track, but he didn't get the laws quite right—that would have to wait for Newton. Ultimately, Descartes argued, matter's behavior derived from the fact that God imparted motion to the corpuscles when he created them. Natural laws then governed how corpuscles moved and behaved when colliding with others.

Given these laws, Descartes deduced how the universe would evolve from its initial state. Circular motion of the parts of matter in the plenum would create whirlpools, or vortices. In fact, the entire universe would be filled with such vortices. As the universe had no limit, it would contain an infinite number (or at least indefinite number) of such vortices. And the eye of each whirlpool storm would contain a star—a sun with a planetary system. For Descartes, as Alexandre Koyré expressed it, "it is perfectly certain that, be the stars far or near, they are, like ourselves and our sun, in the midst of other stars without end."[7]

Descartes denied a true plurality of worlds in the ancient Greek sense. In his system there are not multiple Aristotelian worlds isolated in a boundless space. Rather, Earth and heaven alike are composed of the same matter, which "wholly occupies all the imaginable spaces" where other worlds might exist.[8] So there exists one space, filled with one matter. Yet within that space, whirlpools of that matter created many of the "new" type of world: suns surrounded by planets. "Thus, there are as many different heavens as there are stars," Descartes declared, "and as their number is indefinite, the number of the heavens is the same."[9]

To those who would deny this vast indefinite extension of space, and the multitude of (Copernican) worlds it contains,

Descartes warned that "we should beware of presuming too highly of ourselves." It would be presumptuous for humans to suppose that the universe had limits, absent assurance by "natural reasons or by divine revelation." We should not presume "that we could comprehend by the power of our intellect the ends which God proposed to himself in creating the universe." By "supposing limits . . . of which we have no certain knowledge" to God's creation, we underestimate his power. A modern advocate for the multiverse might interpret Descartes as advising multiverse opponents not to insist that the universe we see—or even what we infer by "the power of our intellect"—must be all that exists.[10]

Descartes is rightly regarded as one of the giants of philosophy. He also aspired to be one of the giants of physics. (Remember, in his day, natural philosophy and physics were essentially synonymous.) Modern physicists do not, however, generally consider Descartes as one of their foremost predecessors in the same league as Galileo and Newton. In fact, most assess Descartes's physical ideas as completely wrongheaded. Steven Weinberg writes that "for someone who claimed to have found the true method for seeking reliable knowledge, it is remarkable how wrong Descartes was about so many aspects of nature."[11]

In some respects, though, it is not so important that Descartes was wrong in the details of his cosmology. It was more meaningful that he viewed the universe in scientific terms, diminishing the role that religious concerns played in the details of physical processes (a major step in making further progress possible). And he was an important figure in initiating the modern quest to understand how the universe was built. "With his physical model, Descartes revolutionized the way of thinking about the structure of the universe," the historian Lucía Ayala has written. Descartes's influence on the rest of the scientific world helped to establish the notion that the universe did in fact comprise a multiplicity of worlds, or Copernican systems in which a central star shone on orbiting planets.

With this new concept of the nature of a world, the modern version of the medieval plurality-of-worlds debate took on a new aspect. It was no longer primarily a theological issue about the creative power of God—it was rather a scientific question about the nature of reality. "Descartes initiated the trend of taking cosmology into consideration strictly within a scientific perspective," Ayala wrote. "Afterward, other authors tackled the origin and evolution of the Universe as a normal topic in astronomy."[12] And as Steven Dick assessed it, "Descartes's vortex cosmology, embracing an infinite universe filled with an infinite number of suns, ingrained the concept of other solar systems in the minds of laymen and natural philosophers alike. The universe would never be quite the same again."[13]

A SPECTACLE FOR THE IMAGINATION

Descartes's impact on scientific thought was substantial. Some of his mathematical achievements—notably, the development of the coordinate system for representing geometrical problems with algebraic equations—were of lasting value. Many of his physical ideas, even if mostly incorrect, served to inspire others to make progress by figuring out why Descartes was wrong. In the realm of cosmology, his influence extended beyond the community of scientists/natural philosophers into the popular consciousness as well. The man responsible for spreading Descartes's word so effectively was the remarkable writer Bernard le Bovier de Fontenelle.

Born in Rouen, France, in 1657 and educated at the Jesuit college there, Fontenelle started a career writing plays and operas, with less than spectacular success. He turned to writing about history and popularizing science, drawing inspiration from the ideas of his countryman Descartes. Fontenelle's writings on the history and philosophy of science, as well as on the new scientific developments of the times, earned him widespread respect from prominent members of the French scientific community. In 1697 he was named permanent secretary of the French academy of sciences; from that position he wrote extensively on science and scientists for the rest of his long life. (He died in 1757, a few weeks short of his 100th birthday.)[14]

As a young man, Fontenelle was a popular visitor at the Parisian salons, where he charmed the ladies with his wit and knowledge of the universe. No doubt such conversations played a part in inspiring his most famous work: *Conversations on the Plurality of Worlds*. He wrote it when he was only in his late 20s; it first appeared in 1686. (He revised it several times throughout his life.) It was a masterpiece of scientific exposition, presented as a dialog between a philosopher and the inquisitive Marquise (given in English by some translations as "Countess").

Fontenelle's *Conversations* explained the latest scientific views on the universe, with special emphasis on the prospects for life on the moon and other planets. "It seems to me that nothing ought to interest us more than to know how the world we inhabit was made, and if other similar worlds exist and have inhabitants also," Fontenelle's narrator opined.[15] Besides the several editions in France, Fontenelle's book was translated into English multiple times and also appeared in German and many other languages. It was widely read throughout Europe by scientists and general readers alike.

"Fontenelle's treatise thus marks a true watershed in the history of the idea of a plurality of worlds," Dick wrote. "By tying the existence of innumerable planets to the image of vortices in explicit terms, the Cartesian cosmology played a central role in extending the idea of a plurality of Earthlike planets to that of a plurality of solar systems."[16]

In the Marquise, Fontenelle created a character in the image of his ideal reader: a woman who knew nothing about the topic but was exceptionally bright, witty, and full of questions. In his preface he describes how he chose such a character by design, both in order to make his work "more enticing" and to provide encouragement for women who, without having formal science education, could still expect to understand scientific ideas if properly pre-sented. "My task is to make it neither too dry for the general public nor too childlike for experts," he wrote. He chose as his topic the new natural philosophy, or physics, for its intrinsic interest. It was something that could pique the curiosity of his readers. "Hap-pily for this subject, the ideas of physics are entertaining in them-selves, and while satisfying to reason at the same time they pro-vide a spectacle for the imagination," he wrote.[17]

In the first of his imaginary after-dinner conversations with the Marquise, Fontenelle's narrator admits to her that he believes the universe contains many worlds, one for each star. "Alas," he says, "I am very sorry to confess my belief that every star could be a world. I would not swear that it is true, but I take it to be true because it gives me pleasure to believe it." Tell me more, the Mar-quise replies: "Since your folly is so pleasant, I will believe of the stars all that you wish me to, provided that I find it pleasing."[18]

Unable to resist her demand to know more, the philosopher dazzles the Marquise with the surprising (to her) news that the Earth moves, rotating on its axis and circling the sun. He describes how Copernicus seized the Earth and sent it far from the center of the universe, to be replaced there by the sun, "to whom this honor is more deserved."[19]

On the second evening, the narrator reveals to the Marquise that the moon is very similar to the Earth and is apparently inhab-ited. He admits that he does not know what such inhabitants might be like, except that no doubt some on the moon would worship the Earth the way some fools on Earth worship the moon. Imagining what the moon's inhabitants might be like is futile. Suppose that moon people tried to imagine what Earth's people might be like. Would they ever conceive of such "a bizarre species of creatures with such foolish passions and wise reflections . . . so much knowl-edge about things nearly useless and so much ignorance of things more important"?[20]

In subsequent conversations, Fontenelle's alter ego takes the Marquise on a tour of the solar system, describing Mercury and Venus, Mars, Jupiter and Saturn, once again proclaiming that they are also probably inhabited. "I see no reason to the contrary," he says. All the planets seem the same—they all receive light from the sun and move in the same manner. "And yet if we are to believe that these vast bodies are not inhabited, I think they were made but to little purpose; why should Nature be so partial, as to except only the Earth?"[21]

Specific features from planet to planet would influence the nature of their inhabitants, however. On Mercury, the sun's heat would drive its inhabitants mad, making Mercury "the lunatic asylum of the universe."[22] Venusians, warmed by the nearness to the sun, would live in a climate favorable for love. Mars offers no remarkable features; it's a boring planet, "not worth the trouble of stopping there."[23] Jupiter is a planet of great beauty, accompanied by its own four little planets (moons), no doubt inhabited also. Jupiter's philosophers would spend all their time making discoveries about their own worlds rather than wondering about Earth, Fontenelle suspects. Saturn's inhabitants would be wise but humorless. As for the sun, Fontenelle says it is clearly unfit for habitation (and if anybody did live there they'd be blinded by the light).

When the Marquise inquires why the little planets—the moons—shouldn't all take their own paths around the sun rather than orbiting their bigger planet masters, Fontenelle seizes the opportunity to explain Descartes's ideas about vortices. "Oh Madame," he said, "if you knew of the vortices of Descartes, these vortices whose name is terrible, though the idea is agreeable, you would not speak as you have."[24] The Marquise laughed and suggested that the vortices should make her head spin, so it would be a good idea to know what they were.

Fontenelle's narrator explains: a vortex, or whirlpool, is a mass of detached pieces of matter all possessing a uniform general motion, all turning round together. Within the "ocean of celestial matter"—Descartes's plenum—the planets all swirl around the sun, which sits motionless (except for its own rotation) in the center of this "great vortex." "But at the same time," the explanation continues, "the planets make little vortexes, imitating that of the sun. Each of them turns around the sun, and rotate around themselves, also turning around a certain amount of celestial matter likewise, which is always ready to follow the motion the planet gives it."[25]

If for some reason a small planet (moon) falls into the vortex of a big planet, it is captured and forced to orbit its new master. Then the big planet, little planet, and the vortex encompassing them whirl together in their orbit around the sun. Thus when the world began, the moon was near enough the Earth to be captured by the Earth's vortex. Four little planets found themselves near giant Jupiter, and so he took all four under his control.

VAST INVISIBLE COUNTRIES

On the fifth evening, the Marquise herself realizes that other stars similarly would entrap planets in their grand vortices, making the universe full of a plurality of worlds. When she asks about the fixed stars, Fontenelle's narrator begins by describing their vast distance from the Earth and sun. "The distance from the sun to the farthest planet, is nothing in comparison of the distance from the sun, or from the Earth, to the fixed stars," he says. And he repeats his belief expressed on the very first night: "All the fixed stars are so many suns."[26]

The Marquise immediately sees where he is going. "I perceive," she says, "where you would carry me; you are going to tell me, that if the fixed stars are so many suns, and our sun the center of a vortex that turns round him, why may not every fixed star be the center of a vortex that turns round the fixed star? Our sun enlightens the planets; why may not every fixed star have planets to which they give light?" To which the narrator happily responds:

"You have said it. And I will not contradict you." Yet even though she anticipated this development in the story, the Marquise reacts in much the way that Kepler did to Bruno's immense universe. Such vastness frightens her. "You have made the Universe so large," she says, "that I know not where I am, or what will become of me." If every star is the center of a vortex as big as ours, then our solar system becomes insignificant, its vast space multiplied many times over, by as many stars as the universe holds. "I protest it is dreadful," she says.[27]

Fontenelle's narrator replies to the contrary: "I think it very pleasant." For Fontenelle, the vastness of the universe is a thing of beauty, offering space to breathe with more freedom, a universe magnificent and glorious. He suggests that the universe may even make new stars from time to time. And without asserting that other worlds (or solar systems) exist beyond even the fixed stars we see, he could not deny the possibility.

"We are arrived at the very roof and top of all the heavens; and to tell you whether there be any stars beyond it, you must have an abler man than I am; you may place worlds there, or no worlds, as you please," he says. "It is enough for me, to have carried your mind as far as you can see with your eyes." And yet "the philosopher's empire" may very well include more "vast invisible countries."[28]

Fontenelle expressed great caution in his opinions, acknowledging that his eloquent depiction of the universe certainly could not be the whole and completely true story. Such caution is an important part of understanding the nature of science. And so Fontenelle's *Conversations* is more than just an engaging explanation of the modern view of the solar system; it is in many ways a sophisticated articulation of scientific philosophy, even if couched in ordinary terms. As he announced in his preface, "I wish to treat philosophy in a manner that is not philosophical."

He compares the scientist to the engineer managing the mechanisms at a theater; science is getting a glimpse of what goes on backstage. "This engineer then is like a philosopher, though the difficulty is greater on the philosopher's part," Fontenelle says, "the machines of the theater being nothing so curious as those of nature, which disposeth her wheels and springs so out of sight, that we have been long a guessing at the movement of the Universe."[29]

Fontenelle clearly believed that even after such a long time guessing, many of the world's wheels and springs remained hidden from the philosopher's view. He derided the idea that the current state of knowledge was all that could be known about the heavens. Humankind's awareness of the cosmos grows with time, as the universe "unfolds" its secrets.

"The world is unfolded by degrees," he tells the Marquise. Ancient scholars believed the Earth's arctic and equatorial zones were too cold or hot to be inhabitable. In Roman times world maps extended only slightly beyond the boundaries of the empire itself. Slowly such ignorance diminished as explorers redrew the maps and discovered that many more parts of the world are populated. Knowledge of these revisions ought to restrain the confidence we place in our current opinions, Fontenelle advises. "The world will unfold itself more to us hereafter."[30]

After all, he explains to the Marquise, philosophy is built on two things: curiosity and poor vision. "Because if you had better eyes than you do, you could see if the stars are suns that illuminate many worlds or not; if on the other hand you were less curious, you wouldn't care to know." In fact, humans are curious. "We want to know more than we see."[31] But we don't always see well; often we see things otherwise than how they really are. So true philosophers do not believe what they see and will constantly conjecture about what they don't see. Our "current opinions" are therefore tentative; we do not yet know all there is to know. "Do we pretend to have discovered all things, or that they have reached a point where nothing can be added?" the narrator asks rhetorically. "For gracious sake, let us consent that there are some things left to be done in the centuries to come."[32]

A SMALL SPECK OF DIRT

Fontenelle's account of a plurality of worlds was not a scientific work. It was written for the public at large. Yet it attracted the attention of scientists as well. One who took particular note was the eminent Dutch mathematician and physicist Christiaan Huygens.

Like Fontenelle, Huygens was a follower of Descartes, at least in a general way, though not in detail. Like Descartes, Huygens advocated the philosophy of a mechanical universe. He believed that things happened because pieces of matter in motion collided, altering that motion and generating the phenomena of the world.

Born in The Hague in 1629, Huygens grew up in a family of well-educated lawyers and diplomats who were well connected with European scientists and other scholars. As a teen he developed wide-ranging interests, from mechanical and carpentry skills to mathematics and music. He attended the University of Leiden, studying math and law, and began physics research, investigating the laws of falling bodies. He studied law further in Breda but found math and natural philosophy a more enticing career choice. His generous father supported Christiaan for the next 15 years or so, allowing him to pursue his research interests without worrying about getting a job.[33]

At first Huygens focused on math, producing (among several other works) an important treatise on probability. He soon also took up astronomy. In league with his brother, Christiaan ground lenses and built fine telescopes, leading to the discovery of Saturn's moon Titan (as well as showing that Saturn's strange bulge was in fact a large ring). Fascinated by time as well as space, Huygens invented the first really useful pendulum clock.

During this time Huygens took occasional trips to Paris, where he met many leading scientists. His works earned him a stellar international reputation. When France formed its Académie Royale des Sciences in 1666, officials persuaded Huygens to become a member and move to Paris. From 1666 to 1681 he lived mostly in

Paris, where he composed several major works, including the famous *Traité de la Lumière,* in which he proposed the wave theory of light. In all his works he promoted the mechanical view of nature that had been so central to the philosophy of Descartes.

Huygens did not, however, buy into all the details of Descartes's physical philosophy, pointing out deficiencies in his analysis of the laws of motion and impact, for example. Huygens did believe in Descartes's vortices and tried to explain gravity with them. But unlike Descartes, Huygens believed that the space between bodies was a void, not Descartes's continuous plenum. Huygens agreed, though, that the so-called fixed stars were just like suns, with their own planets. And like Fontenelle, Huygens believed those planets to be populated.

Toward the end of his life, Huygens expounded his belief in extraterrestrials in *Cosmotheoros,* a short book published in 1698, three years after his death. Huygens considered the existence of multiple worlds to be beyond question. He opened the book by simply stating that that's what the Copernican view of the cosmos implies: "A man that is of *Copernicus*'s opinion, that this Earth of ours is a planet, carried round and enlightened by the Sun, like the rest of them, cannot but sometimes have a fancy, that it's not improbable that the rest of the planets have their dress and furniture, nay and their inhabitants too." Huygens noted that some of the ancient Greeks held the view that there "were an innumerable company of worlds in the stars," and he mentions that Nicholas of Cusa and Giordano Bruno believed that distant stars possessed planets with people as well. But neither Cusa nor Bruno nor "the ingenious French author of the dialogues about *the Plurality of Worlds*" ventures to offer much speculation about the nature of those extraterrestrial beings, leaving open territory for Huygens to conjecture.

Basing his belief on the need to make sense of God's creation, Huygens allowed that so vast a cosmos could not have been made solely for the pleasure of earthlings, whose planet—"this small speck of Dirt"—is "but an inconsiderable point" compared with the extent of the universe. Most of God's creation, an "innumerable multitude of stars," is too far away for human eyes to see and is most likely even beyond the reach of the best telescopes. "Is it such an unreasonable opinion," Huygens asked, "that there are some reasonable creatures who see and admire those glorious bodies at a nearer distance?"[34]

Huygens admitted that such speculations cannot be held true with certainty. But he argued that nevertheless it is reasonable—and logical—to attempt deductions about what cannot be observed based on what is observed. "To reason from what we see and are sure of, to what we cannot, is no false logic," he asserted. "From the nature and circumstances of that planet which we see before our eyes, we may guess at those that are farther distant from us."[35] (Similar reasoning applies today for those who infer an unobservable multiverse from evidence within the visible universe.)

Via a chain of thoroughly articulated logic, Huygens concluded that beings on other worlds must be in many ways similar to earthlings. Senses such as sight, hearing, and touch are most assuredly among their features, as are certain anatomical requirements, such as hands and feet. Huygens endowed his creatures with the power of reason, and morality, suggesting that some planets might even be blessed with "such plenty and affluence" that there would be no crime and everybody would live in perpetual peace. "Perhaps they may not know what anger or hatred are; which we to our cost and misery know too too well."[36]

After describing the planets of the solar system in some detail, Huygens turned to the question of the fixed stars. He argued that the testimony of modern telescopes had removed all reasonable doubt that they are simply a great many suns, "scattered and dispersed all over the immense spaces of the Heaven." He refuted Kepler's contention that the Earth's solar system is surrounded by a special empty space; the space between other stars is just as great as the space between us and any of them. "I must give my vote, with all the greatest philosophers of our age, to have the sun of the same nature with the fixed stars," Huygens declared. Where Kepler despaired of "wandering in immensity," Huygens celebrated the "wonderful and amazing scheme . . . of the magnificent vastness of the universe": "So many suns, so many Earths, and every one of them stocked with so many herbs, trees and animals, and adorned with so many seas and mountains!"[37]

Given the many new stars that telescopes had revealed, Huygens imagined that many times more must be hiding even farther away. Possibly, as Bruno believed, their number was infinite. While Huygens considered Bruno's arguments in favor of infinity to be inconclusive, it seemed impossible to prove the contrary—it was (as Galileo had suggested) a question beyond the capability of humans to answer. The universe itself may be infinite in extent, but what God has placed "beyond the region of the stars, is as much above our knowledge, as it is our habitation."[38]

Huygens was not alone among major scientists of his era in believing in a plurality of worlds within an infinite universe. Isaac Newton (1642–1727) reasoned from his law of universal gravitation that the universe must be infinite as well. He doesn't discuss the issue directly in his masterwork, *Mathematical Principles of Natural Philosophy*, published in 1687. But many years later, in famous letters written to Richard Bentley, Newton outlined his thoughts. If all masses attracted each other (as his law of gravity implied), then in a finite space all the mass would collapse into a great sphere in the center (Aristotle would have been happy with this picture). In an infinite space, though, there would always be mass farther away to counteract the tendency toward collapse. "If the matter was evenly disposed throughout an infinite space, it could never convene into one mass, but some of it would convene into one mass and some into another, so as to make an infinite number of great masses, scattered at great distances from one to another throughout all that infinite space," Newton wrote. In that way, he believed, the sun and stars could have been formed. Why some of that matter coalesced into brightly shining suns while some formed lightless planets, Newton could offer no reason, other than that "the author of the system thought it convenient."[39]

Newton's author was God, of course, to whom Newton attributed the design of the cosmos. The arrangement of planets around a central shining sun, and their motions, could not have arisen by natural causes, Newton averred, but "were impressed by an intelligent agent"—an agent "very well skilled in mechanics and geometry."[40] And that agent, Newton wrote, might well have designed worlds of different types, having created matter particles of different sizes and shapes, in differing amounts in different parts of space, with different densities and forces, "thereby to vary the laws of nature, and make worlds of several sorts in several parts of the universe. At least, I see nothing of contradiction in all this."[41] Newton's universe, then, was very much like the modern notion of the multiverse—infinitely large, containing a possibly infinite number of worlds, of possibly different types, even operating under different laws.

Newton imagined, like Cusa and Bruno before him, a new concept of "universe," altered and enlarged from the ancient and medieval cosmos. And ever since Newton's time, the meaning of "universe" has been in an almost constant state of evolving alteration—beginning with the eigteenth century's realization that there might be more to the cosmos than just the solar system and the fixed stars residing in the Milky Way.

Island Universes

> We have no reason to suppose the Milky Way *really* more extensive
> than the least of these 'nebulae.' Its vast superiority in size is but an
> apparent superiority arising from our position in regard to it—that is
> to say, from our position in its midst. . . . We must picture to
> ourselves a lenticular star-island, or collection of stars;
> our Sun lying excentrically—near the shore of the island.
>
> —EDGAR ALLAN POE, *EUREKA*, 1848

O N T H E F I N A L evening of their series of after-dinner
conversations, Fontenelle's philosopher and the Marquise
turn their discussion and their eyes to the stars, contemplating
their great number. "I can see thousands," says the Marquise.
Whereupon the philosopher informs her that telescopes reveal
vastly greater numbers than those visible to her eye. What's more,
she sees many other stars without knowing it.

"You see that whiteness that's called the Milky Way," the philosopher states. "Can you guess what it is? An infinity of small
stars, invisible to the eyes because they're so small and strewn so
close to one another that they seem to form a continuous whiteness."[1] He suggests that the Milky Way's stars are packed so tightly
together that extraterrestrial pigeons could carry messages from
one to another.

A few decades later, the young philosopher Immanuel Kant
would read Fontenelle's dialogs and come away with a clearer

conception of the Milky Way.[2] In fact, Kant would realize that the sun itself is among the Milky Way's stars. He further understood that the nature of its light indicated the shape of the stellar universe and surmised that other such "universes" exist as well, very far away.

Kant's cosmological deductions, published in 1755, rebooted the debate over the question of a plurality of worlds. Already the definition of "world" had been separated from the previously synonymous "universe." Many worlds—stars surrounded by planets—populated the "universe," as Cusa had speculated and Descartes, Fontenelle, and Huygens had articulated. But by the mid-eighteenth century, this notion of universe as a collection of stellar-planetary "worlds" presented a new question: Was it the whole story? Or did the vast expanse of space contain other such universes? Kant argued that it did. But he was not the first to suggest the possibility. A few years before, the new version of multiple universes had been proposed by the English surveyor and self-taught astronomer Thomas Wright.

AN UNLIMITED PLENUM OF CREATIONS

Wright, born in 1711 at Byers Green Lodge in the county of Durham, was the son of a carpenter. As a teen Thomas left school to apprentice with a clockmaker but avidly read astronomy books in his spare time. His apprenticeship did not turn out to be very successful from an employment standpoint; he eventually studied navigation and supported himself for a while by teaching it to sailors. Along the way he mastered surveying and made connections with some important aristocrats by surveying their estates. His natural pedagogical skills enabled him to make a living through tutoring, lecturing, and scientific writings.[3] Wright did not, however, become a member of the scientific mainstream. And analyses of his writings show that his motivation in doing astronomy was more religious than scientific—his theory of the cosmos was "one stage in a lifelong struggle to integrate scientific knowledge into a satisfying religious vision of the universe."[4] Nevertheless he did inspire a new view of the vastness of the universe, even if not a very accurate one.

Wright was a fan of Newton's physics, which was of course popular in England though not yet universally accepted throughout Europe. In Sweden, for instance, Emanuel Swedenborg theorized about the universe from a Cartesian perspective. In 1734 Swedenborg published an account that recognized the Milky Way as a system of stars circling around a common axis. In other words, the solar system—planets revolving around a common axis through the sun—inhabited a larger "sidereal heaven," the Milky Way (*sidereal* referring to its moving components as stars, not planets). It was a short step for Swedenborg to speculate that other such sidereal heavens occupied the universe.

Apparently independently from Swedenborg, Wright proposed an even more elaborate view of the cosmos, based on Newton rather than Descartes. Newton himself had written very little about cosmology. Apart from his informal correspondence on the subject with Bentley, Newton confined himself to explaining the solar system. Wright went much farther. He believed (as did Newton) that the structure of the cosmos reflected divine architecture and purpose, while also believing that structure could be discerned scientifically. Wright "set for himself the colossal task not simply of constructing a new conception of the physical cosmos but also of integrating into it the domain of the spiritual."[5]

In 1750 Wright presented his new construction of the cosmos in *An Original Theory or New Hypothesis of the Universe.* It consisted of a series of letters; their purpose, Wright stated in his preface, was "an attempt towards solving the phenomena of the Via Lactea [the Milky Way], and in consequence of that solution, the framing of a regular and rational theory of the known universe, not before attempted by any."[6]

In times past, Wright noted, the Earth had reigned in the center of the universe. But it had been put there by human ignorance, not divine wisdom. And neither could the sun be properly placed at the "center of infinity"; there's no reason to think that our sun differs from other stars, so "how can we, with any show of reason, imagine him to be the general center of the whole?" Thanks to the sciences, though, "the scene begins to open to us on all sides, and truths scarce to have been dreamt of . . . invade our senses with a subject too deep for the human understanding, and where our very reason is lost in infinite wonders."[7]

Wright announced that he could deduce the true order of the cosmos given only the premise that the Milky Way is composed of a huge number of small stars. Therefore the "system of the sun" is but a "very minute part of the visible creation." Many other planetary systems exist—Wright derived an estimate of 10 million stars (at least). If each, like the sun, possessed six planets, that equated to 60 million "planetary worlds like ours" in the cosmos. "It appears not only very possible, but highly probable, from what has been said, and from what we can farther demonstrate, that there is as great a multiplicity of worlds, variously dispersed in different parts of the universe, as there are variegated objects in this we live upon."[8]

Wright believed that all these solar systems compose a common sidereal system, the Milky Way. Its milky appearance is an optical effect based on the sun's position with respect to the bulk of its stars. He seems to have believed that the Milky Way's stars (including the sun) occupied the shell of a large sphere.[9] He didn't get the geometry quite right, but the idea of the Milky Way as a stellar system survived and was to be described much more successfully by Kant. (Traditional accounts credited Wright with discerning the true lens-like disk shape of the Milky Way, but twentieth-century scholarship showed otherwise.[10])

"Wright failed to outline a correct (and even reasonable) structure of the Universe," writes Lucía Ayala, "but he did foresee that the Solar System, as traditionally argued in the plurality of worlds by then, was not the proper unit to consider regarding the general scale; instead, it was an element of a larger structure composed of systems with a much bigger size."[11] That is to say, the solar system was just one of many systems collected into a huge galaxy, the Milky Way. And just as the Milky Way galaxy comprised an immense number of stars, so, Wright maintained, did the greater universe encompass countless other galaxies. In consequence of an infinite all-active power, visible creation is full of sidereal systems and planetary worlds. Similarly "the endless immensity is an unlimited plenum of creations not unlike the known Universe," he wrote. That is, the known universe, the Milky Way, is not the only "universe" in the heavens. In support of this notion, Wright cited "the many cloudy spots, just perceivable by us, as far without our starry regions, in which . . . no one Star or particular constituent body can possibly be distinguished."[12] These cloudy spots are the nebulae. Later they would be known as island universes.

AN ABYSS OF A REAL IMMENSITY

Given his religious inclinations, it seemed absurd to Wright to suppose that the Milky Way would be the only stellar system in all of God's creation. God's infinite power demanded the existence of numerous Milky Ways, and the fuzzy nebulae offered (to Wright) obvious candidates to populate his version of the multiverse. Wright's suggestion that the nebulae were in fact distant galaxies marked the beginning of one of the greatest and longest-running debates in the history of astronomy.

Wright's book itself was not directly responsible for the controversy. In fact, his book was not widely distributed in England and did not make much of an impression. But it did attract some attention in Germany, where a lengthy review of it appeared in the Hamburg periodical *Freye Urtheile und Nachrichten* in 1751. Among the readers of that review was the young Immanuel Kant. Kant's development of thoughts stimulated by Wright led to the first truly modern picture of the structure of the universe.

Kant was born in 1724 in Königsberg, then in Prussia but now in Russia (with the name Kaliningrad). His father was the son of immigrants from Scotland. Kant never left Königsberg, attending the university there and then eventually joining the faculty. Although he entered the university to study theology, he grew more interested in math, physics, and astronomy. His focus on philosophy came much later in life (his great work, *Critique of Pure Reason,* was not published until 1781).[13]

Like Wright, Kant became an advocate for Newtonian physics and believed it to be relevant to understanding the universe. Wright's book (or actually the review of Wright's book) inspired Kant to contemplate the physical arrangement of stars in the Milky Way, and what it might look like from a distance. Curiously, the review itself did not mention Wright's "cloudy spots," the nebulae, but Kant on his own realized that they might be other universes.

In ancient times, Ptolemy had observed a handful of such spots, naming them "nebulae" (meaning "cloudy stars"). Galileo noticed many such fuzzy patches of light within the Milky Way when his telescope revealed the countless stars in the milky haze. Some such fuzzballs could be seen even outside the Milky Way's boundaries, Galileo noted. In some nebulae his telescope enabled him to see closely confined stars. In 1612, though, a German astronomer discovered a faint nebula, with no signs of stars within, in the constellation Andromeda. Today we know it as M31, the Andromeda galaxy. In the 1650s, Christiaan Huygens spotted a similar nebula in the constellation Orion. In the decades to follow, other starless nebulae showed themselves as well. Some writers speculated that they might be portals—"openings in the firmament through which one saw into a luminous region beyond . . . the abode of God and the angels."[14] Others supposed that the nebulae were just very dim stars.

But as Kant realized, other stars appeared as mere points of light, while the nebulae were extended. Assuming they were at least as far away as other stars, they must be huge bodies to take up so much space on the sky. Kant "could not believe that such allegedly enormous self-luminous bodies would shine with 'the dullest and feeblest light.'"[15]

Instead, Kant believed that the nebulae must be "systems of stars," much like the Milky Way. His conception of the Milky Way was worked out much more clearly than Wright's. Kant noted that if the sun and Earth sit in the plane of a vast collection of stars, the view toward the plane would take in a swarm of those stars. And that swarm, "on account of the indistinguishable minuteness of their clear points that cannot be severally discerned . . . will present a uniformly whitish glimmer—in a word, a Milky Way." Therefore "our solar world, seeing that this system of the fixed stars is seen from it in the direction of a great circle, is situated in the same great plane and constitutes a system along with the other stars," Kant wrote in *Allgemeine Naturgeschichte und Theorie des Himmels* (*Universal Natural History and Theory of the Heavens*), published in 1755.[16]

Kant then considered what such a system—a disk or lens-shaped mass of stars—would look like from afar. If you viewed such a world of fixed stars at an immense distance from it, you would see it as a circular patch of space if looking straight on, or elliptical if you viewed it obliquely, from the side. And that is exactly how the "nebulous stars" appear. It makes no sense to believe them to be weird, huge, faint stars—it is more natural to view them as "systems of many stars, whose distance presents them in such a narrow space that the light which is individually imperceptible from each of them, reaches us . . . in a uniform pale glimmer." For Kant the conclusion was inescapable: the Milky Way is just one of many galaxies in a vast cosmos, more expansive than humans can conceive. "All this is in perfect harmony with the view that these elliptical figures are just universes and, so to speak, Milky Ways, like those whose constitution we have just unfolded," he wrote.[17]

Kant's insight into the nature of the cosmos is striking; it turned out to be remarkably prescient despite the fact that his scientific expertise was substandard. Several scholars have noted the young

Kant's "quite limited knowledge of science."[18] Still, Kant was one of the few thinkers of his century to grasp the nature of the Milky Way and consider the possible existence of other bodies like it.

Another was the German physicist-astronomer Johann Heinrich Lambert (1728–1777), who published widely on math, philosophy, and meteorology as well as physics and astronomy. Nobody questioned his stellar scientific credentials. In 1761 Lambert published *Cosmologische Briefe,* his description of the universe based on Newtonian physics. Like Kant, Lambert perceived that the Milky Way was a disk-shaped structure. And he imagined that all the stars in the Milky Way possessed, like the sun, planetary companions. "I conclude that the whole system of fixed stars visible to us is not spherical but flat, roughly like a disk whose diameter is many times longer than its thickness," he wrote.[19] Lambert also believed that there might be other Milky Ways—that is, that the Milky Way itself belonged to a larger system of similar objects in analogy to the sun's participation in the Milky Way's system of stars. But he waffled on whether those other Milky Ways should be identified with the nebulae.

Similar waffling can be seen in the changing views of the century's greatest astronomer, William Herschel (1738–1822). An immigrant to England from Germany at age 18, Herschel spent many years as a musician (oboe), including some composing, conducting, and teaching. In the early 1770s he began reading voraciously about astronomy and rented a telescope (not having enough money to buy one). He decided he could build a better telescope himself. With the help of his brother, he built some of the most powerful telescopes in the world, and soon William and his sister Caroline began doing some serious astronomy. In 1781 he discovered a new planet, Uranus, and almost instantly became world famous (even though he thought it was just a comet).[20]

Herschel, apparently with no knowledge of Kant's ideas about the Milky Way's structure, figured out its disk-like nature for himself. But while Kant (and Lambert) had merely theorized, Herschel established the Milky Way's shape by patient observations of the distribution of stars over different patches of the sky.

Herschel developed a particular interest in the nebulae. In December 1781 he received a list of 103 nebulae cataloged by the French astronomer Charles Messier. Herschel began to examine them all. Within most of them his telescope revealed individual stars, and he began to suspect that the nebulae were indeed independent star systems like the Milky Way. Soon Herschel realized there were many more nebulae to be found; within a few years he had identified well over a thousand others. In 1785 he proclaimed to the novelist Frances Burney that he had found "fifteen hundred whole sidereal systems, some of which might well outvie our Milky Way in grandeur."[21]

Later Herschel backed off his commitment to many universes. He had reasoned that the nebulae must be very distant (and hence "universes" separate from the Milky Way) because he could not detect stars in some of them. Presumably a more powerful telescope would have revealed the stars within. But in 1790 he observed such a supposedly distant nebula and found that it was in fact a single star, surrounded by cloudy (nebulous) gas. So Herschel dropped his belief that all the distant nebulae were universes, although he apparently still believed it possible that some of them were.[22]

Even though Herschel lost confidence later, his initial observations certainly strengthened the case for Kant's original ideas about the nature of the nebulae. Kant had clearly envisioned a kind of multiverse, with each "universe" a system of stars such as the Milky Way. Herschel's work seemed to have found many examples of such "island universes"—a phrase often attributed to Kant. But Kant actually did not use that term (despite what unreliable sources like Wikipedia might tell you).

"Contrary to popular opinion, the term 'island universes' never appears as such in [Kant's] the *Theorie des Himmels*," Ayala wrote in 2013.[23] Kant used terms like *Weltgebäuden* and *Weltsystemen* (world systems). When he used the German for island—*Insel*—he meant a real island, like Jamaica, or an uninhabited planet. When Kant said, "These elliptical figures are just universes," the actual term he used in German was *Weltordnungen,* which is literally translated as "world orders."

In truth, the origin of the English phrase "island universe" has long been a mystery. Many accounts cite the opinion of Edwin Hubble, the twentieth-century astronomer who established that the nebulae were, in fact, island universes. Hubble credited that label to the German naturalist and prolific popularizer of science Alexander von Humboldt, famous for his multivolume *Kosmos* (an attempt to survey the entire scope of scientific knowledge). In Volume 3, published in 1850, von Humboldt referred to the distant nebulae as *Weltinseln.* Hubble took that to mean "island universes," and wrote that the phrase was "used in von Humboldt's *Kosmos . . .* presumably for the first time." But as Hubble noted, the first English translations rendered *Weltinseln* literally: "world islands." "The transition to 'island universes' is an obvious step," Hubble wrote, "but the writer [Hubble] has not ascertained the first use of the term."[24] In a modern study of the island universe controversy, historian Robert Smith simply repeats Hubble's account.[25]

In fact, though, von Humboldt was not the first to use the term *Weltinseln* to refer to the nebulae as distant galaxies. And I think I know who first translated it as "island universes." It was an American Civil War general.

MIGHTY UNIVERSES SCATTERED THROUGH SPACE

By the 1850s, the term "island universe" in English had certainly come into use by various writers and speakers. The first use in print recorded by the Oxford English Dictionary, though, was from a book by the American spiritualist Andrew Jackson Davis published in 1867. Davis (who was most concerned with figuring out where in the universe heaven existed) wrote that "island universe" had been suggested to refer to the great distance from the solar system to the fixed stars, indicating that our solar system was the island.

So he didn't even really understand what it meant. John Swett, the superintendent of public instruction in California, also used the phrase "island-universe" in an address to the state teachers' institute in 1867. But astronomers had used the term in public much earlier. Denison Olmsted of Yale, for instance, referred to "island universes" in his Franklin Lectures of 1852. "Island universe" also appeared in Olmsted's book *The Mechanism of the Heavens* (published in 1850), but not in the main text. Rather it appeared in a caption in an appendix of plates illustrating various astronomical objects. And that appendix had been originally published in 1848 in *The Planetary and Stellar Worlds,* by Ormsby MacKnight Mitchel. And that's where the story gets interesting.

Mitchel was born in the Kentucky wilderness in 1809 (or 1810) but grew up in Lebanon, Ohio. With the aid of US Postmaster General John McLean (who had lived in Lebanon), Mitchel gained admission at age 16 to West Point. There he studied astronomy, among other things (such as engineering), graduating in 1829 in the same class as Robert E. Lee. Mitchel continued on at West Point for a while to teach math, then went on active Army duty in Florida. Shortly thereafter he returned to Ohio, where he worked for the railroad (as a civil engineer), studied law, and became a professor of math and astronomy at Cincinnati College.[26]

Mitchel was slight of stature (about 5-foot-6, and he could have boxed as a featherweight), but was a dynamic lecturer, something of an astronomical evangelist. After one lecture in 1842, he announced the nineteenth-century version of a Kickstarter campaign: he wanted to build the country's best astronomical observatory, in Cincinnati. Anyone who contributed (a $25 minimum) would be allowed to look through the new telescope. Money poured in, and Mitchel arranged for a top German firm to build a 12-inch refractor, bigger at the time than any other telescope in America. He found a donor for the land and managed to get the observatory built just in time for the telescope's delivery in January 1845. One slight problem was the lack of funds for any observatory staff. But

Mitchel decided he could run it on his own; his salary from the college was enough, he thought.

But just days before the telescope arrived from Germany, Cincinnati College burned down. Mitchel found himself unemployed with a wife and seven kids. He needed a new business model, quick. He decided on a three-pronged strategy. First he embarked on lecture tours, delivering talks on astronomy to audiences in several cities around the country. Second, he wrote popular books (one of which influenced Richard Proctor, the great American astronomy popularizer of the late nineteenth century). And third, he launched a new astronomy magazine, with a subscription price of $3 per year (payable in advance), that he named *The Sidereal Messenger*. It was in this magazine that the English phrase "island universe" first was used to refer to the distant nebulae.

In Mitchel's 1848 book, "island universe" was used to refer to the Milky Way.[27] But he had used the phrase specifically describing the nebulae in a lecture the year before. As we imagine venturing out into the cosmos, he began, "we pass the confines of our own Universe and sweep on thro' space." Behind us we see our system of stars, one vast cluster. Ahead, "there are some dim hazy spots looming up in the distance. Bring to our aid the telescope—Lo! there burst into view tens of thousands of suns and stars! Here is another Universe burst in upon us, and there is not only one—they are scattered by hundreds and thousands through space. . . . There are more of these mighty Universes scattered through space than there are stars in our system." Imagine what Galileo would have thought, Mitchel mused, if he had been told that future telescopes "would pierce the depths of space and bring out other worlds, other systems, other island universes."[28]

In another lecture, he proclaimed: "Next to the Milky Way, the object is to find what is beyond. Is there anything beyond? Have we reached the end? No; if we were there we should find 10,000 mighty island universes beyond, whose suns must be at least 1,000 millions."[29]

These lectures weren't the first occasions of Mitchel's use of "island universe"—the phrase shows up several times in his *Sidereal Messenger*, the first time in October 1846. In that issue, a long article by the eminent astronomer Johann Heinrich von Mädler refers to "our island universe (since, with reference to the distant nebulae, the entire combination of fixed stars within the Milky Way may be thus designated)."[30] But Mädler had not coined "island universe"; the article was a translation (no doubt by Mitchel) of a paper Mädler had published earlier in 1846 in *Astronomische Nachrichten*. In the original German, Mädler used the term *Weltinsel*—literally "world island"—as von Humboldt did four years later in Volume 3 of his *Kosmos*. So it was Mädler, not von Humboldt, who first referred to nebulae as *Weltinseln*, or world islands, and Mitchel who first rendered that phrase into "island universes."

To be fair, Mädler might have borrowed *Weltinsel* from von Humboldt. It turns out that von Humboldt had, in fact, used *Weltinsel* earlier—in volume 1 of *Kosmos*, published in 1845, the year before Mädler's paper appeared. But in this place von Humboldt used *Weltinsel* just to refer to the Milky Way, not the far-off nebulae. And the English translator rendered it as "cosmical island," not quite the same grandeur as "world island." It is likely that Mädler would have read von Humboldt's book; after all, it was von Humboldt who helped Mädler become an astronomer in the first place. In 1824 von Humboldt introduced Mädler to the wealthy banker Wilhelm Beer, who built an observatory for Mädler to conduct astronomical observations.

As for Mitchel, he used the phrase "island universe" repeatedly in his writings and lectures throughout the 1850s. Then, with the outbreak of the Civil War, he was called to active duty in the Union army as a general. (For some reason the troops nicknamed him "Old Stars.") He commanded the Ohio volunteers for a while and led a raid that took control of Huntsville, Alabama. He was later assigned to a post in South Carolina, where he promptly caught yellow fever and died, on October 30, 1862.

A PROGRESSIVE HIERARCHY OF WORLDS

At the time of Mitchel's death, island universes were much in favor, thanks to observations of the nebula that Huygens had discovered in the constellation Orion. William Parsons, Third Earl of Rosse (1800–1867), using the world's largest reflecting telescope (he built it himself), supposedly resolved individual stars within the Orion cloud. Eventually that observation turned out to be erroneous, but at the time it revived Herschel's original suspicion that a powerful enough telescope would resolve stars in distant nebulae.

In von Humboldt's first use of "world-islands" to refer to the nebulae, he noted that more powerful telescopes might not only resolve the stars within distant nebulae but also bring into view even more distant nebulae in which stars could not be discerned, and so on in endless succession. "Should this not be so," he wrote, "we must, it appears to me, either imagine occupied space to have a limit, or else suppose that the world-islands, to one of which we belong, are so distant from each other that no telescope which may hereafter be invented can ever suffice to reach the opposite shore." Perhaps the most distant visible nebulae would someday be resolved into their individual stars, but von Humboldt questioned whether it is likely that the universe could be so structured, and telescopes could be improved so much, "that in the whole firmament there shall not remain one unresolved nebula."[31]

Here again, then, in a new form, was posed the age-old question of whether the universe contained worlds without end, perhaps extending to infinity, or was singular and finite. Among those also struck by this realization was the poet Edgar Allan Poe, who was well-versed in math and astronomy and followed current events in those fields with great interest.[32] Poe's "prose poem" *Eureka,* published the year before his death in 1849, articulated a remarkably sophisticated (and prescient) view of the nature of the cosmos.

Poe's poem has often been cited as anticipating the modern Big Bang theory of the universe.[33] He also commented directly on the prospect of a multiverse (although he didn't use that term). Poe recognized not only a multitude of clusters of stars (or nebulae)—designating the universe of nebulae as a "cluster of clusters"—but he also alluded to the possibility of multiple such clusters of clusters.

"Telescopic observation, guided by the laws of perspective, enables us to understand that the perceptible Universe exists as *a cluster of clusters, irregularly disposed,*" he wrote. "Have we, or have we not, an analogical right to the inference that this perceptible Universe—that this cluster of clusters—is but one of *a series* of clusters of clusters, the rest of which are invisible through distance?"[34]

Perhaps light from these distant "universes" was too diffuse to be detectable, Poe suggested. Or perhaps they emanated no light at all. Or if they did, their distance might be so vast that "the electric tidings of their presence in Space, have not yet—through the lapsing myriads of years—been enabled to traverse that interval." The right to infer—or imagine—such unseen distant universes depended, Poe said, on the "hardihood of that imagination" claiming such a right. "As an individual, I myself feel impelled to the *fancy*—without daring to call it more—that there *does* exist a *limitless* succession of Universes, more or less similar to that of which we have cognizance."[35]

Nearly a century before Poe's *Eureka,* Kant himself had similarly envisioned a cosmos much vaster than what astronomers' telescopes had up till then revealed. His star-filled nebulae, or "higher universes," he suspected, were part of an even greater, "still more immense system." And he also saw in that the signs of a divine architecture that incorporated worlds (universes) far beyond those that we could now perceive.

"The theory which we have expounded opens up to us a view into the infinite field of creation, and furnishes an idea of the work of God which is in accordance with the infinity of the great Builder of the universe," Kant wrote. It is astounding enough—it "fills the understanding with wonder"—to contemplate the multitude of worlds which fill the Milky Way, "in which the earth, as a grain of sand, is scarcely perceived." And that's not all. Our astonishment is further increased, Kant declared, "when we become aware of the fact that all these immense orders of star-worlds again form but one of a number whose termination we do not know, and which perhaps, like the former, is a system inconceivably vast—and yet again but one member in a new combination of numbers!"[36]

Kant argued that what we can see in the heavens is just the beginning of a "progressive hierarchy of worlds" arranged in systems (planets arranged in a system about a star, stars arranged in a system such as the Milky Way, itself part of a system of island universes)—perhaps an infinite progression. "The first part of this infinite progression enables us already to recognize what must be conjectured of the whole," Kant wrote. "There is here no end but an abyss of a real immensity, in presence of which all the capability of human conception sinks exhausted."[37] That vision hinged, of course, on Kant's being right about the nebulae being island universes. And that question was not settled until the twentieth century.

E Pluribus Universe

> Whether true or false, the hypothesis of external galaxies is certainly a sublime and magnificent one. Instead of a single star-system it presents us with thousands of them. . . . Our conclusions in science must be based on evidence, and not on sentiment. But we may express the hope that this sublime conception may stand the test of further examination.
>
> —A. C. D. CROMMELIN, 1917

EDWIN HUBBLE really did deserve to have a telescope named for him.

Before Hubble, astronomers operated mostly within a limited horizon. Their prime occupation was studying the stars in the Milky Way. Many astronomers believed, much like Kepler, that the space beyond our galaxy was above their pay grade, not a proper subject for science.

"Astronomers in the late 19th century and the very start of the 20th century were very little interested in what we would call the broader universe or its history," says historian of science Robert Smith. "As far as almost all astronomers were concerned, the universe beyond our own limited system of stars was the realm of metaphysics, and working astronomers did not engage in metaphysics."

Other scholars, though, still wondered about the grander cosmos—which was fine with the astronomers. "The infinite universe beyond our stellar system was territory that professional astronomers really were very happy to leave to mathematicians, physicists, philosophers and some popularizers," says Smith.[1]

But Hubble changed all that. He broadened the scope of professional astronomy to encompass the universe as a whole. In the process he revealed a universe vast and dynamic, exceeding the visions of the cosmos imagined by the astronomers of the past. Hubble opened the way to the modern understanding of the expanding, Big Bang–initiated universe that remains essentially the accepted view even today. And the key step in paving the way to comprehending today's cosmos was his validation of the existence of island universes. That led modern cosmologists to redefine the universe as the space containing all those islands. Like a single country comprising many states—*E Pluribus Unum*—the cosmos became a confederation of galaxies. *E Pluribus Universe.*

MAGNIFICENT ISLAND UNIVERSES

When the Earl of Rosse found stars in the Orion nebula, one of the first to celebrate was Ormsby Mitchel. It seemed obvious to Mitchel that Orion was an example of the island universes that Herschel had once believed in before changing his mind. In his lectures and in his magazine *Sidereal Messenger,* Mitchel praised Lord Rosse for his discoveries and devotion to astronomy. "We cannot fail to be struck with admiration when we remember how he devoted his fortune and his energies to the accomplishment of the grand enterprise which has occupied him for years," Mitchel declared. With Lord Rosse's discoveries, the nebulous objects "scattered throughout the heavens" began to reveal their secrets. When Rosse directed his telescope in 1845 toward "the wonderful nebula in the constellation of Orion," for instance, "the stars of which it is composed burst upon the sight for the first time."[2]

Mitchel seems to have believed that all the nebulae were island universes. He cited M13 (the 13th item in Messier's catalog of nebulae), noting that when Edmond Halley discovered it in 1714, he described it as a "little patch of light." In 1764 Messier was unable to say whether stars lurked within it. But when Herschel turned his telescope toward M13, "this object so faintly seen by Messier, burst into ten thousand stars," Mitchel recounted. "It is certainly one of the most magnificent objects in the heavens, and is scarcely ever seen for the first time without exclamations of astonishment. . . . This is doubtless one of the many magnificent 'island universes,' recently revealed by the great instruments of modern times."[3]

But M13 turned out not to be an island universe. And neither is the Orion nebula. Shortly after Mitchel died in 1862, William Huggins dashed the dreams of island universe enthusiasts with a new weapon for conquering the cosmos: spectroscopy.

Huggins, a gifted amateur, was among the first astronomers to realize the power of combining a spectroscope with a telescope. By breaking up the visible light collected via a telescope into its component colors, a spectroscope could reveal an astronomical object's composition.

Born in London in 1824, Huggins was a bright child but had to drop out of school after a battle with smallpox. He still received a fine education from private tutors and intended to attend Cambridge. But his father's poor health required William to focus on running the family business instead. Along the way he had developed an intense interest in science. A gift of a microscope drew his interest to biology, but eventually astronomy became his obsession.[4]

In 1854 Huggins divested himself of the family business and moved to Tulse Hill in the countryside outside London. There he set up an observatory and devoted himself to astronomical research. A few years later, Gustav Kirchhoff (1824–1887) showed how spectroscopy could identify the chemical elements in the sun, and Huggins instantly realized that he could similarly determine the composition of the stars. With the help of a chemist friend, he designed a spectroscope that could be attached to his telescope.

Soon he showed that not all stars were chemically alike—but some seemed very similar to the sun.

In 1864 Huggins analyzed light from the nebula in the constellation Draco and found only a few single-color bright lines—not the full spectrum of colors expected from a large group of numerous stars. He realized that he had solved "the riddle of the nebulae." He could read the answer in the nebula's monochromatic light.

"My surprise was very great, on looking into the small telescope of the spectrum apparatus, to perceive that there was no appearance of a band of coloured light, such as a star would give, but in place of this there were three isolated bright lines only," Huggins later reported. "This observation was sufficient . . . to show that it was not a *group of stars,* but *a true nebula.* A spectrum of this character, so far as our knowledge at present extends, can be produced only by light which has emanated from matter in the *state of gas.*"[5]

Several other nebulae turned out to be gaseous as well. Soon Huggins turned his telescope toward Orion and found the same— it was a gaseous cloud, not far away and full of stars as Lord Rosse had claimed. Points of light thought to be stars must have been just dense gaseous bodies within the more diffuse gaseous nebular cloud, Huggins surmised. "The conclusion is obvious, that the detection in a nebula of minute closely associated points of light . . . can no longer be accepted as a trustworthy proof that the object consists of true stars," he declared. The belief that nebulae must lie far outside the Milky Way (so the "stars" within could not be resolved without powerful telescopes) "must now be given up in reference at least to those of the nebulae the matter of which has been established to be gaseous."[6]

And so Lord Rosse's island universe in Orion turned out to be a mirage. It's one of many instances in the history of astronomy where the claim of a spectacular observation has later been found to be wrong. In this case the confusion is understandable, as astronomer Virginia Trimble has pointed out. "Part of the problem," she writes, "was that the word 'nebula' in fact concealed at least three different sorts of entities—unresolved star clusters in the Milky Way, gas clouds in the Milky Way, and much larger extragalactic assemblages of stars and gas."[7]

Once astronomers accepted Huggins's conclusion that the nebulae were all gaseous, "the existence of island universes became highly questionable," wrote twentieth-century astronomer Jacques Lévy. "At the end of the nineteenth century it was generally believed that the Universe and the Galaxy were one."[8]

But other contributions by Lord Rosse relevant to the island universe debate turned out to be much more important—specifically, his revelation that many of the distant nebulae exhibited a spiral shape.

DISCARDED SPECULATIONS

With weaker telescopes, nebulae looked more or less like fuzzy blobs. But when Lord Rosse examined M51, he could make out twisted arms. His drawing of that nebula looked a bit like a whirlpool, and M51 is known still today as the Whirlpool Galaxy. A few decades later, M51 and other "spiral" nebulae were to become the star players in the island universe drama.

In the meantime, though, Huggins had convinced most astronomers that island universes were a fantasy. Some of the nebulae, like M13, did contain stars, but they turned out to be small clusters of stars (now known as globular clusters) lurking around the Milky Way itself—not vast, distant star systems similar to the Milky Way in size. Other nebulae, Huggins showed, were simply luminous and gaseous. (Herschel had eventually concluded that they contained a "shining fluid" that served as a primordial material from which stars and planets condensed.) And these nebulae also seemed to be members of the Milky Way's own system, not distant galactic Milky Way doppelgängers. On the other hand, the spirals remained a bit of a mystery. They appeared to be scarce in the line of sight looking through the band of the Milky Way, but in other directions a few dozen had been identified. There was no good way to determine their distance, though, so whether they belonged to the Milky Way or resided far away remained unknown.

Still, by the end of the nineteenth century the preponderance of the evidence weighed against the island universe idea. Many astronomers believed instead that the spirals were embryonic solar systems within the Milky Way. Their bright centers were supposedly suns in the making; smaller nebulae in the neighborhood of a big nebula presumably would soon condense into solid planets. A photo of the Andromeda nebula, displayed in 1888 by the astronomer Isaac Roberts (1829–1904), could easily be viewed as strong evidence for the baby solar system hypothesis.

Perhaps even stronger evidence came from Andromeda in the form of a nova appearing in 1885. As novas go, this one seemed especially bright (even though the distance to Andromeda was unknown). In fact, it was nearly as bright as the rest of the nebula itself. It seemed implausible that one star could outshine millions of others—which Andromeda would have to contain if it were indeed an island universe. Nobody then knew about supernovas, which actually can be bright enough to outshine the galaxy they inhabit. So it seemed clear to most astronomers that Andromeda was not an island universe, but rather a new solar system in the process of forming.

Conventional wisdom on the demise of island universes was expressed most acutely by the Irish author Agnes Clerke (1842–1907), a popularizer of astronomy who was highly respected by professionals. In 1890 she wrote that "no competent thinker" could maintain that any nebula was a galaxy on the scale of the Milky Way.[9] In another work, a history of astronomy in the nineteenth century, she belittled the island universe view further. "The conception of the nebulae as remote galaxies," she wrote, had begun by the 1860s "to withdraw into the region of discarded and half-forgotten speculations." She cited an argument from the 1850s by the philosopher and historian of science William Whewell (1794–1866): since some nebulae are seen in the vicinity of star clusters, the two objects are different things, not the same thing seen at different distances. "It becomes impossible to resist the conclusion that both nebular and stellar systems are part of a single scheme," Clerke declared.[10] In other words, there is only one universe, and it's the Milky Way.

Clerke's assessment was shared by others. In 1898 John Ellard Gore (1845–1910), an astronomer and astronomy popularizer, wrote that "the most probable hypothesis seems to be that all the stars, clusters and nebulae, visible in our largest telescopes, form together one vast system, which constitutes our visible universe."[11] Gore did allow the possibility that similar systems might exist in the remote vastness of infinite space. But he insisted that "we live in a limited universe, which is isolated by a dark and starless void from any other universe which may exist in the infinity of space beyond."[12] He thought it possible that the ether—still believed then to be necessary to propagate light—might exist only in galaxies, making it impossible in principle for light from any other universes to reach us through telescopes.

A similar view on the lack of ether at great distances had been expressed in 1877 by the Italian mathematician Giuseppe Barilli, who wrote under the pseudonym Quirico Filopanti. He did believe in the existence of other universes "physically segregated from each other and from ours" but maintained them to be undetectable because no ether was available to transmit light from them to us.[13]

Arthur Berry's well-known history of astronomy, also published in 1898, declared the question of the true nature of the "arrangement of the stars" still unsettled. But he too pointed to the inescapable implication of finding bright (and therefore nearby) stars linked with nebulae. Those nebulae could not be very far away. Added to Huggins's spectroscopic evidence, such observations clearly indicated that the universe comprised a more complicated menagerie of objects than previously suspected. But they still seemed to reside within one galactic mansion. "A good many converging lines of evidence thus point to a greater variety in the arrangement, size, and structure of the bodies with which the telescope makes us acquainted," Berry wrote. "They also indicate the probability that these bodies should be regarded as belonging to a single system . . . rather than to a number of perfectly distinct systems of a simpler type."[14]

But with the turn of the nineteenth century into the twentieth, the island universe story reinvented itself, with the help of some wealthy American philanthropists.

RADIAL VELOCITIES

In January 2017 Robert Smith emphasized the importance of philanthropy as he told of the island universe revival in an engaging talk at the annual April meeting of the American Physical Society. (It was sadly not in a session on time travel; the society sometimes schedules its April meeting in May, or March, or in this case January, for various reasons, but doesn't change the meeting's name).[15] Smith recounted the negative views of island universes—such as those expressed by Clerke, who had repeated her dismissal of island universe fans as incompetent as late as 1905.

But soon after that, Smith noted, philanthropic contributions in support of new, large telescopes, particularly in the American West, led to findings that eventually transformed the one-galaxy universe into today's commodious cosmos, filled with billions and billions of galaxies.

Among the most important of the new astronomical outposts was the Lick Observatory, atop Mount Hamilton near San Jose, California. It was funded by a bequest from James Lick, a land baron reputed at the time of his death in 1876 to be the wealthiest man in California. At Lick, James Keeler (1857–1900) undertook the task of counting the spiral nebulae.

Keeler grew up in Illinois, where as a 12-year-old he witnessed a total eclipse of the sun, which inspired him to build his own telescope. As a college student at Johns Hopkins University, he joined a team that viewed another solar eclipse (in Colorado in 1878). He developed superior spectroscopic skills working at an observatory in Pennsylvania before joining the team at Lick in 1886. Around 1898, when Keeler started seriously studying the spiral nebulae, astronomers knew of only a few dozen. Keeler soon found thousands.[16]

At first, the avalanche of spirals bolstered the case against island universes. If spirals actually did represent solar systems under construction, they obviously must be part of the Milky Way galaxy. And since most of the nebulae did appear to be spirals (as Keeler's work seemed to show), it stood to reason that the rest belonged to the galaxy as well. But Keeler's spirals were soon to testify otherwise—once astronomers figured out how to measure their distance from Earth more accurately.

During the early years of the twentieth century, the renewed interest in spirals stimulated by Keeler led several astronomers to devise methods to estimate their size and distance. Some astronomers even suspected that the Milky Way itself was a spiral, in which case the other spirals must be island universes after all. Important evidence supporting that contention came from Vesto Slipher at Lowell Observatory in Arizona, funded by the wealthy Boston Brahmin (and astronomer himself) Percival Lowell. By 1912 Slipher had begun a systematic effort to quantify how fast the spiral nebulae were flying through space along the line of sight from Earth (their "radial velocity"). Spectroscopy made such measurements possible. Colors of the light emanating from spiral nebulae moving away shifted to the red end of the spectrum. This "redshift" in the light was simply analogous to the lowering in pitch of a receding siren: light from an object flying away would diminish in frequency, or get redder. For nebulae flying toward the Earth, the light's frequency would be higher, or blueshifted.

For most of the spirals that Slipher measured, he found a reddening of the light. But a few (most notably Andromeda) seemed to be moving toward the Earth. In fact, Andromeda appeared to be rushing toward us at 300 kilometers per second, an enormous speed for an object within the galaxy, Slipher realized. Soon he found evidence that some of the spirals moved away from the Earth at even higher velocities. "He would actually start arguing that the spiral nebulae were distant galaxies," Smith said.

During this time several other astronomers began favoring the island universe theory. Arguments and observations relevant to the issue ping-ponged back and forth in talks and papers and letters from one astronomer to another. Among the supporters of island universes was Arthur Eddington (1882–1944) in England, who in 1914 argued that island universes should be "preferred as a working hypothesis." He considered its consequences to be "so helpful as to suggest a distinct probability of its truth."[17] Another astronomer sympathetic to island universes was Andrew Crommelin (1865–1939), who pointed out in 1917 that abundant evidence favored the view that spiral galaxies resided far outside the Milky Way. "It follows that they are of dimensions quite comparable to those of our galaxy," he wrote.[18] Prominent among the island universe skeptics was Harlow Shapley, one of the most famous astronomers of his era, later to become the director of the Harvard Observatory. Most vocal among Shapley's opponents was Heber Curtis of Lick Observatory, who in 1917 composed an assessment of the evidence strongly refuting the view that spirals were solar systems in the making.

In the same year, Shapley argued against island universes in a paper analyzing the brightness of novas in some of the spirals (such as the 1885 nova seen in Andromeda). Shapley drew on measurements of internal velocities in the spirals made by Dutch astronomer Adriaan van Maanen (1884–1936) at Mount Wilson in Southern California, an observatory founded in 1904 by George Ellery Hale (1868–1938) with funding from the Carnegie Institution. Other astronomers believed that the spirals rotated (like hurricanes). But if they are island universes, as Curtis believed, they would be too far away for such rotational motion to be detectable. Measuring such motion would indicate they must be relatively close—part of the Milky Way—and not island universes. Van Maanen investigated and by 1916 claimed to have detected such internal motion in the spiral nebula M101 and others—apparently decisive evidence against the island universe theory.

In his paper, Shapley noted that if the spirals contained a population of normal stars like the Milky Way, the brightness of their novas would place some of them millions of light-years away. In that case, though, van Maanen's measurements could not have revealed internal nebular motion. "Measurable internal proper motions," Shapley wrote, "can not well be harmonized with 'island universes' of whatever size, if they are composed of normal stars."[19]

Island universe advocates did not capitulate, though. The continuing controversy led to what became known as astronomy's Great Debate: a face-off between two astronomers over the existence of island universes on a grand stage at the National Academy of Sciences in Washington, DC, on April 26, 1920.

THE GREAT DEBATE

In one corner stood Curtis. Born in Michigan in 1872, Curtis had an extraordinary aptitude for languages. His education at the University of Michigan included classes in Latin, Greek, and Hebrew, as well as Sanskrit and Assyrian, and he still managed to work in a fair amount of math. In 1894 he got a job teaching Latin and Greek at Napa College in California, where he became interested in astronomy thanks to the availability of the college's telescope. Napa soon merged with The University of the Pacific, which offered Curtis a math and astronomy teaching post. That led to some summers at Lick Observatory, where he learned more about astronomy, transforming his vocational aspirations from language wizard to professional skywatcher. Before long he'd earned an astronomy PhD at the University of Virginia, then returned to California where he began serious research at Lick.[20] By 1917 Curtis's studies of nebulae convinced him that those in the spiral category were island universes, and he became a prominent advocate for that view. He was the obvious choice to defend them in the 1920 debate.

In the other corner, naturally, was Shapley. His path to astronomy was perhaps stranger even than Curtis's—starting out as a journalist while still a teenager, with jobs at newspapers (police beat) in Kansas and Missouri (where he was born in 1885). Shapley determined that it would be best for his career to go to journalism school, just at the time that the University of Missouri had decided to start one. But he arrived before the school was properly organized, and so he took astronomy classes instead. He moved on to graduate school at Princeton where he earned his PhD in 1913. Not long after, he headed for Mount Wilson, where a world-class 60-inch reflecting telescope offered the best view of the heavens available anywhere.[21]

Shapley became the world's expert on globular clusters (like the famous nebula M13), densely packed clumps containing hundreds of thousands of stars. He started out favoring the idea of island universes, but his studies of globular clusters persuaded him otherwise. Shapley also realized that the Earth and sun were not in fact anywhere near the center of the Milky Way. He further concluded that the Milky Way must be much larger than anyone had previously estimated, something like 300,000 light-years across (much bigger than the true value accepted today, in the vicinity of 100,000 light-years). At that size it seemed much more likely that the Milky Way also contained the spirals. (If the Milky Way was that huge, it seemed unthinkable that the universe contained countless other systems of such enormous size.) And van Maanen's measurements clinched the case—for Shapley and many others.

As it turned out, the debate was a dud. There was no back-and-forth—just a paper presented by Shapley and then one by Curtis. Curtis made his case for island universes, but Shapley didn't focus on that issue—instead he offered an elementary discussion of current astronomical issues, emphasizing more his idea of the size of the Milky Way and the solar system's position within it. The real arguments in detail were more thoroughly discussed in the published versions of the papers that appeared later.[22]

Shapley's strongest argument came from van Maanen's observations. If spirals were very distant, the motions he found would have indicated velocities approaching the speed of light. There was no conceivable explanation for speeds that rapid, so the spirals must be nearby. But van Maanen's measurements weren't easy to make. And they turned out to be simply wrong. That became clear after the smoking-gun evidence for island universes emerged from Edwin Hubble's work at Mount Wilson in the early 1920s.

HUBBLE

Hubble was born in Missouri in 1889, moved to Kentucky as a child, and then moved again, attending high school in Chicago. He was both brilliant and athletic (a skilled boxer). After undergraduate school at the University of Chicago, where he studied math and astronomy, Hubble went to Oxford as a Rhodes Scholar and studied law. When he returned to the United States, he briefly considered practicing law in Kentucky, taught high school Spanish for a while in Indiana, and eventually returned to the University of Chicago to earn his astronomy doctorate. Having been influenced by Hale at Chicago, Hubble was prepared to venture to Mount Wilson where Hale was building the best telescope ever, a 100-inch reflector. But World War I was underway, and when the United States entered, Hubble insisted on enlisting in the Army. And so he did not arrive at Mount Wilson until 1919.[23]

In his presentation of the island universe story, Robert Smith emphasized the importance of one of astronomy's greatest discoveries: a way to gauge the distance to Cepheid variable stars, devised by Henrietta Swan Leavitt. Leavitt, born in Massachusetts in 1868, attended Oberlin College in Ohio (where women had long been welcome) and then Radcliffe, where she took a class in astronomy. Despite the problem of a progressively worsening deafness, she excelled at academic work and impressed the director of the Harvard Observatory, Edward Pickering (1846–1919), when she did some volunteer work for him. Later he hired her to oversee his program of mapping the stars with cutting-edge photography and spectroscopy. Leavitt and coworkers measured the brightness of thousands of stars, adding many to the list of Cepheid variables—those that varied in brightness over time.[24]

Such variable stars had been known for over a century (one of the first to be identified, Delta Cephei, gave them their name). But Leavitt analyzed them more thoroughly than anyone had previously and discovered a curious property: they seemed to dim and brighten on a regular schedule. Furthermore, that schedule depended on the star's average brightness. When Leavitt published her paper on that "period-luminosity relationship" in 1908, astronomers suddenly possessed a powerful new tool for accurately measuring distances to astronomical objects. Relatively nearby objects could be determined by their parallax—their change of position on the sky when viewed from one side of the Earth's orbit or the other. That displacement with respect to distant stars enabled a direct distance calculation by simple triangulation. But that method worked only for nearby stars. For more distant objects, parallax was imperceptible.

With Leavitt's discovery, though, a new distance-estimating strategy emerged. If you know a star's intrinsic brightness, you can calculate how far away it is—it's like estimating the distance of a light bulb (of known wattage) by how bright it appears (the apparent brightness diminishes as the square of the distance). Leavitt showed how a Cepheid's brightening-dimming schedule depended on its intrinsic brightness. So once you know the distance of one Cepheid, you can calibrate the brightening-dimming period to deduce the intrinsic brightness of any other Cepheid, from which its distance can then be calculated.[25]

Leavitt's discovery was essential to resolving the great astronomical controversies of the early twentieth century, Smith pointed out during his talk in 2017. "Her discovery of the period-luminosity relationship in Cepheid variable stars is absolutely fundamental in transforming people's ideas about first, our own galactic system and second, providing the means to demonstrate that galaxies do in fact exist," he said.

Indeed, it was Hubble's ability to exploit Leavitt's discovery that enabled him to end the island universe debate. When photographing Andromeda in 1923, Hubble found some brightening stars. At first he thought he'd seen more novas. But he very soon realized that he had actually found Cepheids. Cepheids in Andromeda offered a foolproof way to measure the distance to that nebula. Hubble calculated it to be almost a million light-years from the sun—far beyond even the reach of Shapley's inflated Milky Way. Hubble's calculations provided the death blow to the one-galaxy universe.

Always very cautious, Hubble shared his finding with a few astronomers but decided he needed to verify his results with other Cepheids—both in Andromeda and other spirals—before publicizing his discovery. By late 1924 he'd identified 36 Cepheids in Andromeda and M33, confirming their great distances. His conclusion that the spirals were island universes was inescapable.

Many histories give the impression that the rest of the world first learned of these results on January 1, 1925, when Hubble's paper on the discovery was presented at a meeting of the American Astronomical Society by Henry Norris Russell. (Hubble was unable to attend, so Russell read the paper for him.) But in fact, the Carnegie Institution had already made the results public. Island universe aficionados could read all about them, for instance, in *Science News-Letter* of December 6, 1924: "The spiral nebulae in the sky are universes consisting of uncountable hosts of stars," declared an article headlined "Sky Pinwheels Are Stellar Universes 6,000,000,000,000,000,000 Miles Away." This "dizzying discovery," the report continued, was made by "Dr. Edwin Hubble of the Mt. Wilson Observatory in California."[26]

From its apparent size and measured distance, Hubble estimated the diameter of Andromeda at 45,000 light-years—a substantial underestimate, although clearly indicating Andromeda to be comparable in size to the Milky Way. (In truth, Andromeda is substantially bigger.) "It seems probable that many of the smaller spiral nebulae are still more remote and appear smaller on this account," Hubble commented. "From this point of view the portion of the universe within the range of our investigation consists of vast numbers of stellar galaxies comparable to our own, scattered about through nearly empty space and separated from one another by distances of inconceivable magnitude."[27]

Hubble's assertion about the vast distances separating the nebulae clearly indicated his recognition that the cosmos was much bigger and grander than modern astronomy had up till then grasped. And just a few years later he further established that the space in which those island universes resided was not simply really big, but it was also constantly getting bigger.

BIG BANG

By the end of the 1920s, Hubble, basing his analysis mostly on Slipher's work, was able to show that the radial velocities of redshifted nebulae were directly proportional to their distance from Earth (as measured by Cepheids). In other words, the farther a nebula was from Earth, the faster it appeared to fly away. That relationship formed the observational basis for the expanding universe.

Whether Hubble deserves credit for being the first to demonstrate this distance-velocity relationship is a contentious issue among some historians of astronomy. Other astronomers (including Slipher) had wondered about such a relationship and what it implied. It wasn't obvious, though, just what that relationship was. Perhaps a nebula's recession velocity from Earth was proportional to the square of its distance or some other factor. Astronomers such as Carl Wirtz (1876–1939) and Knut Lundmark (1889–1958) also investigated the possibility of a distance-velocity relationship but did not present a case as compelling as Hubble's.[28]

Meanwhile other investigators began applying Einstein's relatively new general theory of relativity to cosmology. Einstein, of course, led the way with his famous paper of 1917, in which he introduced the cosmological constant (a repulsive energy supposedly needed to keep the universe static, as discussed in chapter 1). He began his paper, titled "Cosmological Considerations on the General Theory of Relativity," by noting that Newton's gravity required a finite island of stars sitting in an infinite space. But such a collection of stars would eventually evaporate into all that empty space. Instead, Einstein proposed, the universe is actually finite. It could still be very big but curved in such a way that it closed on itself, like a sphere. On a large enough scale, the distribution of matter in this universe could be considered uniform, he assumed. (Einstein said it was like viewing the Earth as a smooth sphere for most purposes, even though its terrain is full of bumps on smaller distance scales.) Matter's effect on spacetime curvature would therefore be pretty much constant, and the universe's overall condition would be unchanging—with the repulsive force of the cosmological constant preventing gravitational collapse.

Dutch mathematician Willem de Sitter (1872–1934) also found solutions to Einstein's equations—for a universe containing no matter. But it wasn't immediately clear how to draw conclusions from de Sitter's theory relevant to the real universe. A major (but unappreciated) advance came in papers published in 1922 and 1924 by the Russian mathematician and meteorologist Aleksandr Friedmann (1888–1925). He rejected Einstein's insistence on a static universe and showed how general relativity could allow for a dynamic universe, either contracting or expanding. Friedmann died in 1925, however, and the value of his contribution was recognized only after Hubble showed the universe to be expanding.

If anyone else does deserve more credit than Hubble for that realization, it would be Georges Lemaître (1894–1966), a Belgian priest (and astrophysicist). In 1927 Lemaître also derived equations describing an expanding universe and estimated the universe's expansion rate based on radial velocity data. But his paper was published obscurely in a Belgian journal, in French (and an English translation a few years later omitted crucial passages). In any case it was Hubble's 1929 paper that did get the astronomical community's attention, persuading other scientists of the dynamic nature of the cosmos (or at least to take the idea seriously).

So, regardless of who actually figured it out first, it was Hubble who alerted the world at large that the universe was getting bigger. A second paper, published in 1931 by Hubble and coauthored with the observer Milton Humason (1891–1972), further cemented the case for the expansion of the universe—one of the most profound discoveries about nature since Copernicus demoted the Earth to solar-satellite status.[29]

Ironically, Hubble was not very vigorous about asserting his discovery. He was too cautious to say anything more in his 1929 paper than that the rapid recession of distant galaxies might validate one aspect of de Sitter's work. "The velocity-distance relation may represent the de Sitter effect . . . [including] a general tendency of material particles to scatter," Hubble wrote in his 1929 paper.[30] A few years later, in lectures (published in book form in 1936 as *The Realm of the Nebulae*), Hubble acknowledged that "the velocity-distance relation is considered as the observational basis for theories of an expanding universe."[31] But he continued to warn that other explanations might be possible, remaining skeptical that the universe really expands.

In any event, no matter who gets credit for what, the period 1923–1931 witnessed a dramatic revolution in astronomers' image of the universe they studied, on the order of the original revolution launched by Copernicus. From the nineteenth century's universe as a lone galaxy in a static space, the cosmos became the home of countless such galaxies, in a space that was constantly expanding. Eventually cosmologists gathered the clues from theory and observations into a mathematical framework later to be known as the Big Bang theory—the story of a universe that began small, hot, and dense, and then expanded to generate an ever-growing space giving birth to galaxies, stars, planets, and people.

For a period from the late 1940s to the mid-1960s, an opposing view of the universe's biography captured serious attention. Known as the steady-state theory, it postulated an eternal cosmos, expanding, to be sure, but maintained in a uniform state by the constant creation of new matter from empty space. Evidence favoring the Big Bang view continued to accumulate, though. The 1964 discovery of a cold bath of microwaves permeating all of space—the remnant afterglow of the initial Big Bang explosion—persuaded all but a few of the Big Bang's validity.[32] By the 1970s it seemed that humankind had finally come to a complete and sure understanding of what the universe is: a grand expanding spacetime arena, housing billions and billions of galaxies. And that was all there was—one universe, undivided, with a beginning in time billions of years ago. Indeed, this universe contained a multiplicity of worlds: myriad island universes. But once again the definition of universe simply morphed into a new form, returning the number of worlds to one.

So that seemed to be the end of the centuries-long argument that the atomists had initiated in ancient Greece and had been revived among medieval philosophers by the edict of 1277.

Ah, but the atomists had another trick up their sleeve. Mess with the atomists, and they'll come back at you. With quantum mechanics.

Many Quantum Worlds

The framework of quantum theory allows, and seems to demand,
that we envisage ourselves within a multiverse of a qualitatively
new kind. The traditional "cosmological" Multiverse considers
that there might be physical realms inaccessible to us due
to their separation in space-time. The quantum Multiverse
arises from entities that occupy the same space-time.

—FRANK WILCZEK, "MULTIVERSALITY," 2013

T'S STARDATE 5693.2, and the starship *Enterprise* is en-
snared in the Tholian Web. Captain Kirk is lost in space somewhere, although he seems to reappear briefly at regular intervals.
Mr. Spock determines that the captain has slipped into another
universe. "We exist in a universe which coexists with a multitude
of others in the same physical space," Spock explains. "At certain
brief periods of time, an area of their space overlaps an area of
ours." In the twenty-third century, it seems, science has established
the existence of the multiverse.[1]

It's a curious kind of multiverse, though, unlike any debated
by medieval scholars or even the multiple bubbles of spacetime
inferred from inflationary cosmology. Spock is talking about a
quantum multiverse. He does not explain it further.

But jump ahead to Stardate 47391.2, when the next generation's *Enterprise* is boldly going where no one has gone before.
Lieutenant Commander Data, an android (humanlike robot, not

a phone) tells the rest of the story. It seems that Lieutenant Whorf, the Klingon who serves as the *Enterprise*'s security officer, has been experiencing memory problems. "Things are changing," Whorf complains. "It is as if events and circumstances continue to change from moment to moment, but I am the only one who seems to be aware of it."

Whorf offers his own nonquantum explanation: "Someone is playing a trick on me." But Data runs some tests and discovers that the "quantum signature" in Whorf's cellular RNA reveals him to be in the "wrong" universe. Whorf has arrived on a different *Enterprise* after traveling through a quantum fissure in the spacetime continuum. At such a fissure, Data observes, many "quantum realities" intersect.

"For any event, there is an infinite number of possible outcomes," Data points out. "Our choices determine which outcomes will follow. But there is a theory in quantum physics that all possibilities that can happen do happen in alternate quantum realities." And Whorf, it seems, has been "shifting from one reality to another."[2]

Science fiction writers often make up such weird theories to "explain" their hard-to-believe plot twists. But in this case Data's theory of alternate quantum realities came straight from real science. It was introduced to the world in a physics PhD dissertation written in the 1950s by a freethinking genius named Hugh Everett III. The ideas expressed in that thesis came to be known as the Many Worlds Interpretation of quantum mechanics. Historian of physics Max Jammer referred to it as the "multiuniverse" theory. "The multiuniverse theory is undoubtedly one of the most daring and most ambitious theories ever constructed in the history of science," Jammer wrote. "In the scope of its conceptions it is virtually unique."[3]

Its scope did in fact exceed anything previously contemplated, whether by the Greek atomists, medieval philosophers, or the early twentieth century's astronomers and cosmologists. Not only do multiple universes exist, Everett's interpretation suggested, but also all the possible variations of all physical events actually transpire in an uncountably many such universes. These universes do not exist far away, independently of one another, but coexist, each new one the offspring of another, like branches sprouting from a very complex tree. New branches constantly split away from previous ones as different possible outcomes of quantum processes occur.

At least that is how the many-worlds idea is commonly expressed. It's hard to say, though, exactly what Everett had in mind. And subsequent approaches based on Everett's idea have further obscured his meaning. That shouldn't be surprising, because just as quantum physics implies multiple realities, quantum mechanics has spawned multiple interpretations of what it really means. Everett's interpretation is just one of many interpretations of quantum mechanics. But it has become one of the most popular. And it is responsible for an unprecedented and unexpected new chapter in the story of the human quest to comprehend existence, describing an entirely novel version of the multiverse.

BIRTH OF THE QUANTUM

Quantum physics originated in the early years of the twentieth century from multiple lines of research aimed at understanding radiation (especially light) and explaining the atom. German physicist Max Planck fired the first shot in the quantum revolution in December 1900. In pondering experimental data on the colors of light emitted by a "black body" (think of it as a hot oven), Planck deduced that the radiation must be emitted in packets rather than a continuous flow. He called such a packet a quantum. He knew it was a big deal, but Planck did not pursue the most revolutionary ramifications of his idea. Albert Einstein, much more daring, proposed in 1905 that quanta of light travel through space as particles. (Planck had assumed that in flight the packets merged into the familiar electromagnetic waves.) Einstein's particles (much later designated "photons") explained the photoelectric effect, but few other physicists bought the idea.[4]

A much more positive reception toward quantum ideas came with the model of the atom proposed by the Danish physicist Niels Bohr in 1913. Bohr's quantum atom consisted of electrons in well-defined orbits around a central positively charged nucleus. An atom absorbed or emitted a quantum of light when an electron hopped from one orbit to another, Bohr explained. His model helped make sense of Ernest Rutherford's discovery (just two years earlier) that most of an atom is empty space. Almost all of an atom's mass, Rutherford calculated, resided in a very dense nucleus at its center. Lightweight electrons buzzed about at a great distance. If you expand an atom to the size of a football stadium, the nucleus would be a golf ball on the 50-yard-line with electrons represented by fleas hopping among the cheapest seats.

Bohr's original model successfully explained experimental data for the simplest atom, hydrogen. But it failed for more complicated atoms. By the mid-1920s, though, the German physicist Werner Heisenberg, one of Bohr's protégés, developed more sophisticated mathematics that made quantum ideas applicable generally to the mechanisms of the atomic and subatomic world—producing the analog to the classical science of mechanics. "Quantum mechanics" soon became the accepted mathematical framework for understanding light, matter, and energy.

Making sense of quantum mechanics was difficult, though. Heisenberg did not himself fully understand the mathematical methods he created. He had unknowingly devised a form of matrix algebra—news to Heisenberg, but well-known to his professor at the University of Göttingen, Max Born. With help from Pascual Jordan, a German mathematical physicist, Heisenberg and Born worked out the quantum math more rigorously. At about the same time, Austrian physicist Erwin Schrödinger developed totally different quantum math in the form of an equation that described electrons as waves rather than particles. Fortunately for the future of physics, Schrödinger's wave-based math and Heisenberg's particle-based math turned out to be mathematically equivalent. They made the same predictions for the outcomes of experiments. But conceptually the gulf between the two views seemed like a chasm worthy of an Evel Knievel rocket stunt.

On the conceptual level, Bohr addressed the problem by formulating his principle of complementarity. Electrons might sometimes present themselves as waves, on other occasions as particles. It depended entirely on what type of experiment you performed, Bohr said. On the practical side, Born realized that Schrödinger's wave math could be interpreted as a way of forecasting the probabilities for various experimental outcomes. Schrödinger's mathematical "wave function" allowed calculating probabilities for such things as an electron's position.

But there was a slight problem. Heisenberg had shown in 1927 that electrons did not actually occupy any specific positions. Bohr's old idea of well-defined orbits did not really apply to electrons. Rather than running around the nucleus in a narrow ring, an electron hovered about in something more like a negatively charged cloud. If you conducted an experiment to pinpoint an electron—make it run into a phosphorescent screen, for instance—you could localize it at the spot of the impact. But until then, the electron did not occupy a particular location or travel in any well-defined path (a consequence of what became known as Heisenberg's uncertainty principle). An electron existed in multiple locations at once, technically known as a superposition of states.[5]

It did not take long for everybody to realize that what was true for electrons was true for other particles, and therefore for matter in general. Reality itself, at least at the subatomic level, could no longer be viewed as one single chain of cause-and-effect events. Multiple possibilities for the configurations of matter existed in the math.

Only one such possibility manifests itself to human senses, though. When you walk into a room containing a single chair, you don't see several shadowy fuzz blobs sort of in the shape of a chair floating about. You see one chair in one spot. In the world of ordinary life, you do not encounter a multiplicity of coexisting realities. Observations, or measurements, somehow destroy all but one of the quantum possibilities. Just how and why that happened came to be known as the "quantum measurement" problem.

Bohr attempted to dismiss that mystery by distinguishing quantum processes from everyday experience. Whatever is going on at the quantum level is inaccessible to human perception. Of all the many possible outcomes of a quantum process, only one becomes "real," because the attempt to observe or measure that process can record only one outcome. The result of an observation or measurement must be communicated in ordinary human language; there is no way in principle to describe what goes on in the fuzzy realm of quantum multiplicity before a measurement is made. "It is wrong to think that the task of physics is to find out how nature *is*," Bohr liked to say. "Physics concerns what we can *say* about nature."[6]

Bohr's philosophy smoothed the way for most physicists to use quantum mechanics in practice without worrying too much about its deeper implications. On a technical level, procedures for applying quantum math were worked out in detail by Paul Dirac in England and by the Hungarian mathematician John von Neumann. In brief, a quantum system is described mathematically by a wave function (or state vector) that specifies the "location" of a system in a multidimensional mathematical space (known as Hilbert space). The wave function allows multiple possible outcomes for a measurement performed on the system but does not tell you which of those results you will get. You can, however, use the wave function to calculate precise probabilities for obtaining the various possible results. Those probabilities change over time in a well-defined way as long as the system remains isolated from outside influences.

So far, there is nothing to complain about. But when the system is disturbed, say by a physicist performing an observation, the situation changes drastically. The observation yields a definite result—a value for whatever is being measured from among the many possibilities represented in the wave function. In other words, when the measurement is made, the wave function "collapses"— many possibilities disappear, leaving just the one that is measured.

This perspective poses no problems for anybody who just wants to use quantum mechanics for making experimental predictions. But it nevertheless baffled those who wanted to know what was really going on at the quantum level. It seemed inconceivable that nature at its roots should operate with such a split personality. Among those most troubled about it all was Everett, after he entered Princeton University as a physics graduate student in 1953.

HUGH EVERETT III

Everett was born in Washington, DC, in 1930. He grew up nearby in Maryland; his father had a military background and his mother was a writer. His parents divorced when Hugh was five. He attended a Catholic high school and then enrolled at Catholic University of America (in DC), majoring in chemistry but excelling most demonstrably in math. He was also psychologically aloof and socially inept. "Everett was an independent, intense, driven young man," wrote John Archibald Wheeler, Everett's thesis adviser at Princeton.[7]

Soon after arriving at Princeton, Everett became enamored with game theory—the mathematics of competitive (or sometimes cooperative) interactions—an emerging field of which Princeton was the acknowledged center. But he also took classes in quantum mechanics. In one class, taught by Nobel laureate-to-be Eugene Wigner, Everett encountered the philosophical conundrum posed by the quantum measurement problem.

Around that time, Bohr delivered a lecture in Princeton on complementarity. Everett did not find Bohr's views logically satisfying and so began thinking about developing a new approach of his own. His analysis of the quantum problem would become his PhD thesis.

In need of a thesis adviser, Everett presented his thoughts to Wheeler. Wheeler had been at Princeton since the 1930s; he played prominent roles in the Manhattan project to build the atomic bomb and in the subsequent hydrogen bomb project. Wheeler was also a disciple of Bohr's, having spent time at Bohr's institute in Copenhagen and then collaborating with Bohr in 1939 on the paper that explained the theory of nuclear fission. Wheeler had just embarked on a plan to revitalize the study of Einstein's general theory of relativity, and a key question at the time was how general relativity, which successfully described gravity, could incorporate quantum mechanics. So Wheeler expressed great interest in Everett's idea, even if at first it seemed "barely comprehensible."[8]

Everett argued that the "wave collapse" view of measurement presented paradoxes. Let's say you're at work in your lab, which contains a quantum system in a box. (It could be Schrödinger's cat if you don't like cats, but that doesn't matter.) You make your measurement and record the result in your notebook. The wave function describing your quantum system has therefore collapsed. But suppose your lab is just one small room in a bigger building. Your boss thinks of your lab as a box containing a quantum system that includes you (in the role of the cat). To your boss, the wave function of your lab contains many possible outcomes for the result you recorded in your notebook. If your boss waits a week before entering your lab and looking at what you recorded, quantum physics says your result was not "real" until then. For a week, then, you believed you had recorded a solidly real result but actually hadn't (from the point of view of the world outside your lab).

You see the problem. Either you are just a fuzzy shadow deluding yourself into thinking you had obtained a real result, or your boss's cherished quantum mechanics was wrong for a whole week about what was going on in your lab. Everett found this paradox intolerable. "It is now clear that the interpretation of quantum mechanics with which we began is untenable if we are to consider a universe containing more than one observer," he wrote.[9]

For one thing (and ultimately, a major thing), how could quantum mechanics be applied to the entire universe? There is no boss "outside" the universe to make a measurement about what's going on inside and select only one of the possibilities. And so, Everett decided, the answer was obvious: all the possibilities are real. If quantum mechanics describes a superposition of different possibilities, one observer's measurement does not eliminate all but one of them—the others still exist. "It is therefore improper to attribute any less validity or 'reality' to any element of a super-position than any other element," Everett wrote. "All elements of a superposition must be regarded as simultaneously existing."[10]

While Everett disliked complementarity, he did not view his approach as completely contradicting Bohr. Complementarity could fit into Everett's universe easily enough—it just wasn't the whole story. In fact, Everett suggested that he was offering a way to reconcile the conflicting views of Bohr and Einstein.[11] Einstein disliked the probabilistic nature of quantum theory's predictions—on many occasions, he declared that "God did not play dice" with the universe. And Einstein found Bohr's insistence that an obser-vation somehow created reality was an affront to the notion of an objective reality independent of any observer. Einstein some-times illustrated his objection by arguing that the moon existed whether anybody was looking at it or not. In April 1954, speaking at Wheeler's relativity seminar at Princeton, Einstein made his point another way. "When a person such as a mouse observes the universe, does that change the state of the universe?"[12]

Everett's answer to this question was that the mouse did nothing to the universe, but the universe did play a trick on the mouse. It's the mouse who somehow splits into different realities. Suppose its quantum system is a mousetrap, in a superposition of both set and not-set. An attempt to nibble the cheese determines whether the trap goes off or not. At the first bite, two versions of the mouse emerge—one less hungry, one dead. "From the standpoint of our theory, *it is not so much the system which is affected by an observation as the observer, who becomes correlated to the system*," Everett wrote.[13] An observer making a measurement interacts with the system being measured and is therefore connected (or correlated) to that system. But within the entire universe there are many possible correlations. While the observer is correlated with a specific result for the nearby system, the rest of the universe is in a quantum state that retains a superposition of other possibilities.

"From the present viewpoint all elements of this superposition are equally 'real,'" Everett asserted. "Only the observer state has changed, so as to become correlated with the state of the near system and . . . with that of the remote system also. The mouse does not affect the universe—only the mouse is affected."[14]

It is this aspect of Everett's theory that suggests a splitting of the universe into branches—in one branch the mouse gets the cheese, in another the trap snaps. As Everett described it at a conference a few years later, you can imagine a series of measurements made by an observer. "With each observation, the state of the observer splits into a number of states, one for each possible outcome," Everett said. "Thus the state of the observer is a constantly branching tree." Each branch describes the history of observational results that the observer has obtained, and the statistical frequencies of the results will be "precisely what is predicted by ordinary quantum mechanics."[15]

In that way, Everett believed he had addressed both Bohr's and Einstein's concerns. "Our theory in a certain sense bridges the positions of Einstein and Bohr," Everett wrote.[16] As Einstein believed, the universe evolves in a purely deterministic way—God does not play dice—yet the results for any particular observer will observe a statistical pattern predicted by the quantum math, as Bohr insisted they would.

One physicist who had seen Everett's paper before publication raised an objection: an observer feels no splitting into different states as one possibility becomes actual. Everett replied in a note (added in proof to the published version of his thesis) that there is actually no transition from possible to actual. "From the viewpoint of the theory *all* elements of a superposition (all 'branches') are 'actual,' none any more 'real' than the rest," he wrote. "It is unnecessary to suppose that all but one are somehow destroyed, since all the separate elements of a superposition individually obey the wave equation."[17] Nobody feels any splitting because the branches exist independently; nobody feels the rotation of the Earth, either, but that doesn't mean the Earth is motionless, Everett pointed out.

Everett's thesis was a compromise. Wheeler aided him in cutting down a long version to the shorter version published in 1957 in *Reviews of Modern Physics*. Partly Wheeler wanted to ensure smoother sailing for Everett with the other members of the PhD committee, and partly Wheeler wanted to limit Everett's criticism of Bohr, who thoroughly rejected the whole idea. Wheeler also felt it necessary to write a short paper accompanying Everett's publication "to make his thesis more digestible to his other committee members."[18]

Everett's concept "does demand a totally new view of the foundational character of physics," Wheeler wrote. But he emphasized that the conclusions reached from Everett's view implied that experimental results would correspond exactly with the conventional (Bohr's) view. Bohr's view of measurement could be seen as just a special case of Everett's more comprehensive theory, and in fact Everett's math might even provide "the possibility of a closed mathematical model for complementarity."[19]

Wheeler called Everett's theory the "relative state" formulation of quantum mechanics, and that was the label used in the published version of Everett's thesis. Everett titled the original long version of the thesis "The Theory of the Universal Wave Function," emphasizing the application of quantum math to the universe as a whole.

While Everett referred to branches of this universal wave function, he did not use phrases like "parallel universes" or "many worlds." He seemed rather to believe in one universe that just so happened to encompass multiple branches of reality that sprouted in various directions as physical systems interacted. At a conference in 1962, he did informally use the term *worlds* in response to some questions. "Each individual branch looks like a perfectly respectable world where definite things have happened," he said. "Each separate element of the superposition will obey the same laws independent of the presence or absence of one another."[20]

Nobody paid much attention to Everett's thesis during the 1960s (except perhaps for Mr. Spock). That's not unusual for bold new ideas; scientists often spend years writing follow-up papers and promoting their work at conferences before others take notice. But Everett did not pursue the academic path. Despite encouragement from Wheeler to continue working on the quantum issue, Everett opted for a job at the Pentagon. He became a key player in Cold War obsessions with analyzing military nuclear options, drawing on his expertise with game theory. His personal life eventually deteriorated; as one biographical account put it, "Everett was addicted to alcohol, food, tobacco, and sex"; his obesity and heavy smoking led to death by heart attack in 1982 at age 51.[21]

His interpretation of quantum mechanics did not die with him, though. In 1970 an article in a quasipopular physics journal had jettisoned the technicalities and accentuated the drama of Everett's thesis as a theory of multiple universes. Ever since, Everett's view has been a prominent part of the quantum (and multiverse) conversation.

MANY WORLDS

Among the physicists who perused the draft of Everett's thesis was Bryce DeWitt (1923–2004), then at the University of North

Carolina. After reading Everett's paper, DeWitt was "simultaneously delighted and shocked," he recalled decades later. "This was something fresh. There had been so many extremely dull papers."[22]

But he did have one concern about the observer "splitting"—it was DeWitt who complained to Everett that nobody "feels" themselves splitting into multiple branches of the universe. But Everett persuaded DeWitt that his theory posed no logical inconsistencies. And DeWitt became Everett's champion.

In the late 1960s, DeWitt's graduate student Neill Graham investigated Everett's work and developed it further for a doctoral dissertation. DeWitt decided it was time for Everett's idea to be more widely known and submitted an article to a relatively accessible forum, *Physics Today*. ("Accessible" in this case meant readable for trained physicists who did not specialize in the particular topic of the article.) DeWitt's *Physics Today* article, titled "Quantum Mechanics and Reality," appeared in September 1970, introducing Everett's views to the world at large.

DeWitt's focus, he wrote at the outset, would be on a proposal that "pictures the universe as continually splitting into a multiplicity of mutually unobservable but equally real worlds." It leads to a "bizarre world view," DeWitt acknowledged, but nevertheless "it may be the most satisfying answer yet advanced" for why measurement always produced only one of the many possible results contained in the quantum math.[23]

In somewhat flowery language (for a physicist), DeWitt articulated the consequences of adopting Everett's view. If quantum math accurately describes the universe, then "this universe is constantly splitting into a stupendous number of branches, all resulting from the measurementlike interactions between its myriads of components," DeWitt wrote. "Every quantum transition taking place on every star, in every galaxy, in every remote corner of the universe is splitting our local world on earth into myriads of copies of itself." He reiterated that no observer existed outside the universe who could ascertain which of the many branches represented reality. So instead, "all are equally real," he wrote, and "each branch corresponds to a possible universe-as-we-actually-see-it."[24]

While DeWitt alluded to these branches as "maverick worlds," he did not use the phrase "many-worlds interpretation" in his *Physics Today* article (he referred to the Everett-Wheeler-Graham or EWG proposal). But in 1973, DeWitt and Graham edited a book with the title *The Many-Worlds Interpretation of Quantum Mechanics,* and that is how Everett's idea became generally labeled (although it is sometimes called the "Many Universes Interpretation"). Wheeler, however—although he wrote supportively of Everett's idea at first—did not like the idea of parallel universes. "More than I could swallow," Wheeler wrote.[25] Later Wheeler disavowed the Everett theory, complaining that "it creates too great a load of metaphysical baggage."[26]

DeWitt, on the other hand, insisted that simplicity should be sought in the equations, not the outcomes that the equations produced. Many-worlds math is as trim as it can be, he said, yet still contains that stupendous number of universes. "They're there, in the equations," DeWitt told me when I interviewed him in 1987. Later, at a symposium in 2002, he continued to insist that the other worlds were real and might even be observed someday. "Observed—that's a bad word, perhaps," he added. "We will know that they are there."[27]

While DeWitt's article and book made Everett's ideas well-known among physicists, there was no immediate rush to dump Bohr's approach, known as the Copenhagen interpretation. But over the years, as Bohr's disciples died off, many younger physicists gravitated to something closer to Everett's view. When I talked to DeWitt for a story marking the 30th anniversary of Everett's thesis, DeWitt expressed confidence that it would become the prevailing interpretation. "It's not dying out," he said. "I think the younger physicists are coming to accept it more and more. They haven't been brainwashed" by the traditional perspective.[28]

DeWitt noted that physicists exploring cosmology and applying quantum mechanics to the early cosmos could not avoid taking Everett seriously. "In order to discuss these things physicists have to write down a wave function for the whole universe," DeWitt said. "And as soon as you do that you're completely in the Everett camp anyway, and they know it."[29]

One Everett enthusiast was David Deutsch, a young physicist at Oxford University. He emphasized the importance of wave interference as evidence for the other universes. If you send quantum particles (say photons, or electrons) through a barrier with two slits, a detector screen on the other side will record alternating bands of light and darkness—an interference pattern. That's clear evidence that those "particles" are behaving like (or actually are) waves. Ordinary physics can explain such interference patterns. Where the waves meet in phase (crest corresponding to crest, trough to trough), they boost each other and make an impact on the detector screen. Where waves meet out of phase, they cancel out. The resulting alternating bands demonstrate the wave nature of quantum particles. But the interference pattern is produced even if the particles pass through the two-slit barrier one particle at a time. It's hard to see how a particle could interfere with itself. Deutsch contended that the interference was caused by a particle/wave in another universe.

Deutsch argued that this view could be verified with a quantum computer. Back then, though, in the 1980s, few people knew what he was talking about. The idea of a computer designed to exploit the weirdness of quantum physics had been proposed by Paul Benioff and further discussed by Richard Feynman, but purely in hypothetical terms. Instead of computing with ordinary bits, represented by either 1 or 0, a quantum computer can calculate using quantum bits in a state of both 0 and 1 at the same time. Deutsch declared that such a computer could solve problems much more rapidly than an ordinary computer—by computing in multiple universes.

Today, of course, quantum computing is a lively field of research and several such computers have been built—although on too small a scale to be very useful just yet. They do, however, illustrate the quantum computing process that Deutsch offered as proof of many universes. Other physicists do not share his interpretation, though. I emailed Charles Bennett of IBM, a quantum computing pioneer, about this point back in 1994. He did not believe that quantum computers supported Everett's interpretation. All major interpretations of quantum mechanics predict the same results when you make a measurement of a quantum system, even if your system is a quantum computer, Bennett pointed out. A successful quantum computer "is no more a rigorous proof of the many worlds interpretation than the ability of light to propagate in a vacuum is a proof of the existence of the luminiferous ether."[30]

QUANTUM DECOHERENCE

While quantum computers don't prove the Many Worlds Interpretation, they don't refute it either. And Everett's interpretation has in fact gained in popularity over the years. In particular it lives on, in modified form, within several other approaches to coping with the quantum measurement problem.

One development during the 1980s became especially popular for dealing with the paradoxes posed by quantum weirdness: the study of a phenomenon known as quantum decoherence. Decoherence occurs when a quantum system interacts with its environment. The many possibilities in the coherent quantum wave function "decohere," leaving one reality. While the idea wasn't totally new, interest in it was sparked by a paper in 1985 by Erich Joos and H. Dieter Zeh. The decoherence approach was subsequently promoted extensively by Wojciech Zurek, a Polish-American theoretical physicist working at the Los Alamos National Laboratory in New Mexico, where I interviewed him in 1993.

Zurek saw his approach as a way to bridge the chasm between Everett's and Bohr's interpretations, with decoherence as the chief pillar holding up the bridge. In Zurek's view the many Everett possible realities that an observer might encounter really do exist—but only very briefly. Interactions of an object with its environment establish it in a well-defined position because soon only one position is consistent with all the interactions. A chair in a room seems to sit in one spot because only one spot is consistent with the trajectories of all the air molecules and light particles that have bounced off it.[31]

Zurek showed how to calculate how fast this decoherence process happens. It depends on size, mass, and temperature. For a very small object, on short distance scales, at very low temperatures, decoherence can take a while—that's why electrons in atoms can seem so weird. But for a large, heavy object at room temperature—say, a chair—decoherence acts rapidly. A chair assumes its well-defined location faster than a light beam can travel the width of an atom. That explains why nobody ever sees a chair in a superposition of positions.

It's true, Zurek noted, that some atoms still may have interacted with the chair when it was in another possible place. But humans can perceive only a fraction of all of reality, so the odds strongly favor seeing only one of the many possibilities—and that would be the one most robustly recorded by interactions with the environment.

"When we perceive a system, we intercept a tiny fraction of the environment that has interacted with that system," Zurek explained at a conference in 2002. "If this fraction is so tiny, this means the environment can reveal to us only the observables of the system which were recorded redundantly." So as far as people are concerned, the remaining possibilities lose their meaning, making Everett's alternate branches of the universe irrelevant. "We have no chance of seeing these other parallel universes," Zurek said.

He didn't think they should count as real. "To me," he said, "reality is the touchable everyday reality."[32]

A related approach incorporating the decoherence idea was developed in the 1980s and 1990s by Murray Gell-Mann and his collaborator James Hartle. Decoherence alone, Gell-Mann and Hartle argued, is not enough to resolve the quantum-measurement dilemma. Decoherence allows the environment to play the role of

Murray Gell-Mann

James Hartle

an observer to condense specific results out of the fog of quantum possibilities. But if you want quantum physics to apply to the whole universe (and you do), there is no "environment" outside the universe to appeal to. Somehow quantum physics needs to apply to the whole enchilada, an entirely closed system, such as the universe itself.

In pursuit of that goal, Gell-Mann and Hartle combined decoherence with a view called "consistent histories," developed in the 1980s by Robert Griffiths of Carnegie Mellon University. Griffiths proposed that even though quantum systems can start out with many possibilities, as time goes by only a few combinations of those possibilities would continue to make logical sense. As quantum processes occur, soon only some "realities" are logically consistent with all the events that have come before.

When I wrote about this in my book *The Bit and the Pendulum*, I compared it to the movie version of the board game *Clue* (Tim Curry, Martin Mull, Paramount Pictures, 1985). Throughout the film, various clues are revealed about who might have killed Mr. Boddy. But the ending of the movie—revealing the actual killer—depended on which theater you went to. The theatrical release came in three versions. In one the killer was Mr. Green, in the hall, with the revolver. In another Mrs. Peacock committed the murder, and in yet another the maid did it. But all three endings were consistent with all the events and clues that had been divulged earlier.

Gell-Mann and Hartle argued that quantum reality behaved something like that. But they noted that a consistent history at the level of human observations could in fact contain many subhistories with subtle variations. One apparently single history is actually a set of consistent histories; if you looked at a small enough scale, you could see minor differences. But if you ignored the microscopic level, you saw one reality.

In explaining this idea to me, Gell-Mann used the example of a horse race. During a day at the races, a lot goes on that has nothing to do with the orders of finish. Stray atoms kicked out of the dirt by the horses' hooves won't influence who wins, places, and shows. At the end of a day (say with eight races, each with ten horses), one list of outcomes (from 100 million possible permutations) gets posted online at the *Daily Racing Form* website. But that list ignores all sorts of unnoticed variations (like which molecules went where after getting kicked out of the dirt). Race fans at the track that day did not engage in one single history, but a set of very similar histories all consistent with the races' outcomes.

Life in general is like that, in Gell-Mann's view. Differences from history to history within a set are at too minuscule a scale to detect. So they don't affect the reality observed at the macroscopic, or "classical," level—the predictable world of ordinary experience. We don't notice a lot of the fuzzy possible quantum variations, just an apparently classical world where events all seem rock-solid real and are consistent with what has happened earlier.

This approach allows for a description of the emergence of classical (or "quasiclassical") reality in a closed quantum system, such as the universe. Think of the universe as a big box, containing the particles and fields governed by the laws of physics, Gell-Mann and Hartle suggested in one of their later papers. "Everything is contained within the box, galaxies, planets, observers and observed, measured subsystems, any apparatus that measures them, and, in particular, any human observers including us," Gell-Mann and Hartle wrote.[33] Quantum processes within the box must generate concrete reality without an outside observer. Consistent histories within that box must tell logical stories—in Gell-Mann and Hartle's terminology, "narrative coarse grainings."

Coarse graining is a familiar idea in physics. Physicists study matter all the time without worrying about what's going on at the level of each individual atom or molecule, just as in everyday life road maps do not depict potholes. In a similar way, all histories within a set of consistent quantum histories produce the same large-scale observable events.

Gell-Mann and Hartle referred to their approach as "decoherent histories" and called a set of consistent histories, where reality seems so solid, a quasiclassical realm. They considered their work as "an attempt at extension, clarification and completion of the Everett interpretation."[34]

Not all of Everett's branches would produce quasiclassical realms suitable for habitation. In a quasiclassical realm, the level of coarse graining is the most refined graining possible while preserving the features of classical physics. Such "adaptive" coarse graining occurs only in some branches—whether a branch contains things like planets and stars depends on the specific features of the immediately preceding branch. "That way the quasiclassical realms can be a property of our universe and not just our choice," Gell-Mann and Hartle noted.[35]

In our branch, one protostellar gas cloud became denser than others, leading to the birth of the sun. In general, where matter is denser, decoherence happens more rapidly, leading to the construction of various objects within environments; those environments preserve the history of those objects. In this way adaptive coarse graining generates narrative histories (or sets of histories) that tell a consistent story of how the universe evolves. "Narrative sets of histories . . . allow the construction of an environment," Gell-Mann and Hartle wrote. "They therefore put the notion of environment in its proper place as a consequence of a narrative coarse graining and not as a separate postulate of quantum mechanics."[36]

Gell-Mann and Hartle believed that their approach generalized quantum mechanics in a way that made it applicable to the entire universe. There was no need for an observer to make quantum possibilities real. Gell-Mann told me he did not consider it a refutation of Bohr's Copenhagen interpretation, though—just a more broadly applicable generalization.

"It's not wrong, but the Copenhagen interpretation is extremely special," Gell-Mann said. "But nowadays for quantum cosmology we want a quantum mechanics that applies generally, not just to a physicist repeating a physics experiment over and over again in the laboratory."[37]

In my last interview with Gell-Mann, in 2009, he reiterated that point about the Copenhagen interpretation. "It's correct in a sense, but it can't be general—it can't be the deep way to look at quantum mechanics." But he also insisted that his approach did not really favor the many-universes theory as it had been popularized by DeWitt. "Everett came up with an approach which I would say is a forerunner to ours—it's not the same exactly, but he goes a large part of the way toward what we have," Gell-Mann said. "He was along the right lines, but then Bryce DeWitt came in with this many-worlds idea, that you had to have a world for every alternative history of the universe. I don't know why you needed a world for each one."[38]

Many other variations on Everett's idea have been explored by quantum theorists during the twenty-first century. In its various forms it has become one of the most prominent of the many competing interpretations of quantum theory. But Everett's interpretation is certainly not generally accepted—no one interpretation is. Advocates of the view that the quantum wave function is "ontic"—referring directly to reality—tend to support some version of the Everett interpretation. But many experts prefer the view that the wave function is "epistemic"—a useful calculational tool representing our knowledge about a quantum system.[39] Some recent attempts to cope with the quantum problem explicitly deny Everett's multiverse altogether, such as the "transactional" interpretation, which holds that the alternative possibilities are "real" merely in the sense of Aristotelian potential—only some of them become actual.[40] And the many-worlds view is very much at odds with an interpretation known as QBism, which contends that a quantum state is math used by scientists to make predictions, not an objective set of multiple realities.[41]

Still, in discussions about the possibility of parallel universes, Everett's "many worlds" are almost always mentioned. On occasion it has even been suggested that Everett's many worlds somehow correspond to the multiple universes produced by inflation, but that is a minority view. In any case, Everett's branches do not really belong in the same category as the multiple universes contemplated throughout history with each successive redefinition of what a universe is. The atomists' dreams of many universes have been fulfilled not by quantum mechanics, but by inflationary cosmology.

Anthropic Cosmology

It's very easy to imagine laws of physics that would give you a
boring universe. But we seem to live in one which has a very rich
cosmology. It has laws that allow for a rich chemistry and the . . .
elements that you would need for life. So at some intuitive level,
some very imprecise and intuitive level, it does seem that the
laws of physics that we see are unusually suitable for life.

—JOE POLCHINSKI, 2006

THE ANCIENT GREEK physician Galen was very seriously interested in astronomy.

He avidly read the works of Hipparchus, the greatest of the
Greek astronomers before Ptolemy. Like most scholars in his times,
Galen (AD 130–200) believed that the heavenly bodies, especially
the sun and moon, exerted influences on the ground. Those influences extended to matters of life, death, and disease; the sun's light
causes growth, Galen observed, and the moon's phases affected the
course of his patients' illnesses.

Galen also believed that the hand of a creator had specified the
arrangement of the heavenly bodies as well as designing the parts
of living bodies. Though the sun is "grand and most beautiful"
while a foot is a "small, ignoble part of an animal," nonetheless "the
same skill has been employed in locating both sun and foot," Galen
observed. "In both cases the art of the Creator is equally great." Sun
and foot "could not be located in any better place."[1]

His logic seemed solid. It was hard to see where you would rather have your feet. And a sun much farther or closer would make life unbearable, either freezingly chilly or too tropically hot. Galen therefore viewed the sun's comfortable distance from the Earth as evidence for divine design in the cosmos.

But today's cosmologists—some of them at least—interpret the illustration differently. They see it as an argument favoring the existence of the multiverse.

ANTHROPIC ANXIETY

After 1998, when astronomers discovered that the universe is expanding faster and faster, the multiverse went mainstream. In much the way that the Condemnation of 1277 opened the debate over the plurality of worlds, the reports in 1998 of cosmic acceleration licensed scientists to advocate for a multiplicity of universes. Once again, science's consensus conception of the scope of reality had to be reconsidered. A debate supposedly settled so many times before reignited itself.

Cosmic acceleration did not merely rekindle the debate about multiple universes, though. It provided explosive new ammunition for supporters of a physics heresy known as the anthropic principle.

During the last few decades of the twentieth century, several scientists had noted that certain features of the cosmos seemed suspicious.[2] Nobody could explain why the universe is just the age it is, why the forces of nature are just the strength they are, why the mass of a neutron is just a little bit more than the mass of a proton. Yet changing any of those features even slightly would yield a universe inhospitable to complexity—in particular biological complexity. In that case no humans, or anybody else, would be around to wonder about such things. Apparently humankind inhabits a Goldilocks universe, with everything "just right" for life. This suspicious suitability of the universe for human habitation led some scientists to explain it by invoking the anthropic principle— that the existence of life requires the universe's fundamental features to be something very close to their observed values. Or as Steven Weinberg once phrased it: "One of the reasons that things are the way they are is that if they weren't that way, we wouldn't be here to ask the question."[3]

To many scientists, though, the anthropic principle was about as popular as root canals, telemarketing, and Internet trolls. Some even referred to *anthropic* as the A-word. And as an explanation for various features of the universe, most physicists gave the A-word an F. "Anthropic reasoning is defeatist, and it's dangerous," physicist Joe Lykken of Fermilab once told me. "Frankly, any idiot can come up with anthropic explanations of things."[4]

Much of the dislike for the anthropic principle stemmed from a belief that it smacked of intelligent design, à la Galen's perfectly placed feet and sun. If the universe's properties had been finely tuned to permit life to thrive, then some designer must have done the tuning. Hard-core physicists insistent on keeping God out of physics denied the need for such anthropic explanations, promising that a soon-to-be articulated "theory of everything" would specify the universe's blueprint in sufficient detail to account for every cosmic property. But then the plot twisted: the secret weapon (code name, superstring theory) against the A-word turned the debate upside down. Rather than doing away with anthropic cosmology, superstring theory provided one of the strongest arguments for it (and for the multiverse), shocking almost everybody. Except Andrei Linde.

THE SELF-REPRODUCING UNIVERSE

Even before the discovery of the accelerating universe, inflationary cosmology had infected a small cadre of cosmological investigators with multiverse fever. Linde in particular had realized that inflation implied a new view of the standard Big Bang birth of the universe. "If you consider any kinds of these inflationary models, then you have a possibility of self-reproduction of the universe," he told me in 1991. "At the moment it doesn't seem that we have proof that there was a single big bang from which everything started. . . . It's much more probable that the thing which we now call the big bang was not the first bang. I would say that what we are seeing now perhaps was not *the* big bang but was one in a sequence of bangs." That sequence would continue forever, Linde believed, in a process called eternal inflation.[5]

As you'll recall from chapter 1, Linde was one of the early investigators of inflation theory, initially proposed in 1980 by Alan Guth. Guth's main idea—a brief instant of super-rapid expansion—held up over time, but the technical details didn't. Attempts to repair inflation's technical flaws led to various new versions of inflationary cosmology throughout the 1980s. "New inflation," the initial successor to Guth's original version, implied the ongoing creation of multiple universes, work by Alex Vilenkin of Tufts University showed. But new inflation ultimately failed to pass other tests. More successful was Linde's "chaotic inflation" scenario, first proposed in 1983. It eliminated new inflation's flaws while retaining its multiple universes. In an energy field permeating the chaos of primordial space, Linde calculated, some spacetime bubbles would have inflated to enormous size, forming regions like our universe. Later Linde was able to show that chaotic inflation also allowed eternal inflation, a term he first used in two papers in 1986. He gave one of them the self-explanatory title "Eternally Existing Self-Reproducing Chaotic Inflationary Universe."

In describing his ideas, Linde alluded to a well-known conversation between Einstein and his assistant Ernst Strauss about whether God had any freedom of choice in creating the universe with its particular suite of properties. "What really interests me is whether God could have created the world any differently," Einstein told Strauss.[6] (Of course, Einstein was not arguing for religion, but was speaking metaphorically—he posed the question in the context of whether more than one "logically consistent" universe was possible.)

Linde's investigation of chaotic inflation led him to answer in the affirmative. Not only was it possible to create a different universe, it was impossible not to. Chaotic inflation would not produce just one universe, but many. In the pervasive field of energy from which our universe burst into existence, quantum effects "lead to an infinite process of creation and self-reproduction of inflationary parts of the universe," Linde wrote. Not only did many bubbles burst into existence from primordial chaos, but new small bubbles could even sprout off from big ones. "One may say therefore that not only could God create the universe differently, but in His wisdom He created a universe which has been unceasingly producing different universes of all possible types."[7]

At first, Linde referred to the many bubbles produced by eternal inflation as "mini-universes." (Guth prefers the label "pocket universes.") No matter how our universe began (if it actually had a beginning), "it now contains an exponentially large (or even infinite) number of mini-universes," Linde wrote. "*All* types of mini-universes in which inflation is possible should be produced during the expansion of the universe, and it is unreasonable to expect that our domain is the only possible one or the best one."[8]

He didn't say "best for what." But the key point was that the mini-universes would not be identical. Within each bubble, the processes that produced nature's particles and forces would have created a unique work of cosmic art. Each bubble would be like a snowflake. All snowflakes form under the guidance of the same basic physical laws, but minute variations during the process of their formation ensure that they will all differ from one another. New universes would similarly form adhering to fundamental laws, but quirks in their development could produce differences in the strengths of their forces and even properties of space itself. "It seems absolutely improbable that all domains contained in our exponentially large universe are *of the same type*," Linde wrote.

So our universe is not, as Einstein might have hoped, the only possible one, Linde contended. If chaotic inflation occurred, our bubble is just one of countless many. "In our opinion," Linde wrote, "the modification of the point of view on the global structure of the universe and on our place in the world is one of the most important consequences of the development of the inflationary-universe scenario."[9]

As with Everett's many quantum universes, Linde's multiverse was mostly met with indifference. But a few physicists, such as Vilenkin, took the new multiverse idea seriously and did substantial research on it. And one prominent advocate for the multiverse eventually promoted the idea widely: the astronomer royal of Great Britain, Martin Rees. "Our entire universe may be just one element—one atom, as it were—in an infinite ensemble: a cosmic archipelago," Rees wrote in a popular book. "This new concept," he wrote, "is, potentially, as drastic an enlargement of our cosmic perspective as the shift from pre-Copernican ideas to the realization that the Earth is orbiting a typical star on the edge of . . . just one galaxy among countless others."[10]

When Rees's book was published in 1997, his was a minority view. Soon thereafter, though, his enthusiasm received some unanticipated justification.

ANTHROPIC SUCCESS

With the discovery of the accelerating universe in 1998, many physicists quickly realized the need to reconsider their aversion to two discredited ideas: Einstein's cosmological constant and the anthropic principle. Accelerating expansion implied the existence of some sort of energy field permeating the vacuum of space and exerting a repulsive effect, driving spacetime to expand. Such a vacuum energy sounded just like the cosmological constant (so named because its strength should be constant at all points in space) that Einstein had proposed eight decades earlier. And if Einstein's cosmological constant truly existed, then the respectability of anthropic reasoning suddenly became enhanced, too—thanks largely to an important paper by Weinberg that appeared in 1987.[11]

If the energy driving the universe to accelerate really is Einstein's cosmological constant, its discovery validated a prediction that Weinberg made by applying anthropic reasoning.

Most physicists at the time believed that Einstein's cosmological constant wasn't a terrible idea. Quantum physics allowed (or perhaps even required) some vacuum energy that would behave just as Einstein envisioned, with negative pressure that would drive space apart. Yet few believed this field really existed, because if it did, we wouldn't. Theoretical calculations indicated that Einstein's vacuum energy density should have bloated the universe much too rapidly to allow stars or planets to form. Physicists commonly presumed that some unknown mechanisms contrived to cancel out the vacuum energy, leaving its value at precisely zero. Weinberg wasn't so sure.

Weinberg is widely viewed by other physicists as the embodiment of gravitas and one of science's deepest intellects. He was not known for advocating ill-founded speculation. And he was among the last physicists anyone would expect to favor anything like the existence of parallel universes. Yet in his 1987 paper, published in *Physical Review Letters,* Weinberg found a rationale for believing in the cosmological constant, with important implications for the multiverse debate.

Weinberg calculated that the cosmological constant did not have to be precisely zero to allow people to exist. A value a little bigger than zero, if not too much bigger, would permit a cosmic expansion rate slow enough to allow stars and galaxies to form. If you want a universe with life, you need some sort of "bound gravitational system" that, once formed, would no longer be affected by the cosmological expansion, Weinberg pointed out. Some initial clump of matter in the early universe must have the opportunity to grow into something big before the vacuum energy carried away the matter it could grow from.

Weinberg's calculations indicated that some amount of vacuum energy remained compatible with the formation of galaxies big enough to allow stars to form and then explode, providing the chemical elements needed for life. Besides that, he noticed, a value for the cosmological constant consistent with anthropic reasoning might even help solve some other perplexing problems. For one thing, inflation theory predicted a "flat" universe, meaning omega, the measure of matter density, must be very nearly equal to 1. But there just didn't seem to be enough matter in the universe to make omega equal to 1. Measurements suggested that omega was equal to about 0.2 (that is, the matter density was only 20 percent of the amount needed to make the universe flat). But energy (including vacuum energy) also contributed to the overall geometry of space, Weinberg pointed out. With a value for the cosmological constant within the anthropic limit, the combined mass-energy density of space might still make omega equal to 1. Second, the oldest stars seemed older than the expansion rate of the universe allowed them to be. A cosmological constant would have altered the past expansion rate of the universe enough to make the ages of the stars and the universe compatible.

Two years later, in a review paper, Weinberg explicitly concluded that anthropic reasoning implied the existence of some vacuum energy (a cosmological constant). "If it is only anthropic constraints that keep the effective cosmological constant within empirical limits, then this constant should be rather large, large enough to show up before long in astronomical observations," he wrote.[12] And that's exactly what happened in 1998, with the first reports of cosmic acceleration—implying that space is subsumed by a vacuum energy with properties just those expected of the cosmological constant.

In 2003 Weinberg spoke at a conference in Cleveland on the future of cosmology. During a panel discussion, he mentioned that his 1989 paper had contained the prophecy forecasting the cosmological constant's discovery. "I don't want to exaggerate the importance of this remark," he said, but paused briefly and then added, "That's not true—I do." It was an important point, he said, that anthropic reasoning had pointed in the right direction. "The anthropic argument had a chance to fail," Weinberg said, "and it didn't take it."[13]

At a public lecture the night before the panel session, a member of the audience asked Weinberg's views on the anthropic principle. "There is a context in which it makes perfect sense," Weinberg responded. If the Earth happened to be the only planet in the universe, it would be very odd for it to find itself at just the right distance from the sun for the temperature to allow water to be liquid. You might have to consider it as evidence for a benign creator, as Galen had suggested. And Weinberg said he would have agreed with Galen if the Earth was the only planet that existed.

"If the Earth is the only planet in the universe, I would have no other explanation of why we're so lucky," Weinberg admitted. "But now we know that there are a lot of planets, and a fair fraction of the stars in our galaxy have planets . . . and so it's not at all surprising that there should be one or maybe several or maybe many on which the conditions are suitable for life. And it's only on those that there are people who are asking the question."[14]

It seems natural enough to extend the reasoning about the Earth's convenient temperature to the lucky amount of vacuum energy in the known universe. Which Weinberg did during the next day's panel discussion.

"What you need for this kind of explanation is many parts of the universe with varying values of the cosmological constant," he said. Those "many parts" might even be other universes, of course—possibly the many bubbles of chaotic inflation, Linde's mini-universes. Or, Weinberg emphasized, the "many parts" might be components of the wave function of the universe—Hugh Everett's (or Bryce DeWitt's) "many worlds." Anthropic reasoning may imply a multiverse, but it doesn't tell you just what kind of multiverse. Just as in the Middle Ages, when philosophers discussed many sorts of other worlds, the modern multiverse can take many different forms (to be discussed more fully in chapter 14).

A MATHEMATICAL DEMOCRACY

Multiple forms of the multiverse were on the agenda in Princeton, New Jersey, in 2002 at a conference celebrating John Wheeler's 90th birthday.[15] In a talk at that conference, the nonconformist cosmologist Max Tegmark outlined a multiverse categorization that has become widely adopted, sorting the various versions into four levels. Tegmark's Level I includes just the regions of our own space beyond the reach of our telescopes. If, as the best evidence indicates, our universe is infinite, then countless regions of space exist beyond our sight, dwarfing the comparatively tiny sphere of space (billions of trillions of miles across) that humankind's telescopes can probe. Although such space extends outward without end, each region could contain only a finite number of atoms. So in an unlimited multiplicity of spaces, every possible arrangement of those atoms must sooner or later occur in some realm somewhere. In fact, each arrangement would occur over and over again in regions more and more distant. And consequently, somewhere, someone identical to you is reading an identical copy of this book.

Somehow that seems troubling. But it is possibly comforting to realize that there is no prospect of running into any of your doppelgängers. Tegmark calculated that the nearest copy of any of us would be at a distance (in miles) computed by multiplying 10 times itself 100 trillion trillion times.

Tegmark's Level I multiverse does not really seem much like a multiverse, though. It's the same universe we live in, just very big. Much more interesting is Tegmark's Level II multiverse, in which the fundamental laws of physics apply, but manifest themselves in different local or "effective" laws from place to place. It's Tegmark's Level II multiverse that intrigues people like Linde and Guth, because cosmic inflation provides a mechanism for creating many spacetime bubbles with such differing local laws and properties. In the inflationary picture, bubbles emerge from the false vacuum—the inflaton field—and grow into independent universes.[16] In some versions of this view, black holes would form in the inflationary process; in a sufficiently large black hole a new "baby" universe would grow, at first connected to its parent universe by a wormhole, a sort of cosmic umbilical cord. Under some conditions, the umbilical cord eventually evaporates or "pinches off," leaving the new universe a complete orphan.[17]

In a sense, all those members of the multiverse belong to the same "universe," but they exist as completely independent space-time realms. Nature's physical constants could differ from bubble to bubble; even the number of spatial dimensions could be different, as could the strength of vacuum energy. Such differences in local physics are the key to the anthropic implications of a Level II multiverse.

Tegmark identified his Level III multiverse with the many worlds of quantum mechanics. He didn't stop there, adding a Level IV: universes defined by "other mathematical structures." Our universe is the physical representation of a body of mathematical equations, he noted. But he could see no reason why other logically consistent mathematical structures couldn't exist and serve as the basis for a reality of their own. "My own personal guess," he said, "is that you have a mathematical democracy—they're all equal—and any structure that exists mathematically exists physically, too."[18]

Linde saw the mathematical question a little differently. For him, the question was not why a particular mathematical structure describes nature's particles and forces, but rather why we live in a place where math works so well. Inflation provides the answer, he said in his talk at the Wheeler celebration. Inflation provides bubbles with a variety of particles and forces; in some bubbles, mathematics can efficiently describe how those forces and particles interact to allow stable, complex structures to form. In realms where such stability is impossible, math would not be so useful. "We live in places—which are fortunately quite abundant in this multiuniverse picture—where stability is possible," Linde said. "Mathematics is efficient in the places where we can live."[19]

Despite the success of such anthropic reasoning, many prominent physicists remained unmoved, their attitude ranging from skeptical to hostile. At the Cleveland meeting in 2003, Nobel laureate-to-be David Gross expressed a common sentiment: "I hate the anthropic principle," he said. "My feeling is that anthropic reasoning is kind of a virus. It seems once you get the bug, you can't get rid of it."[20]

Gross and others contended that accepting the anthropic principle amounted to giving up on what they saw as the purpose of physics: to find a theory that specified everything. By everything, they meant the fundamental features of the cosmos: the underlying laws and physical constants from which the details of how the universe worked could be deduced. The correct theory of everything should eventually be able to explain the value of the cosmological constant from underlying physical principles and mathematical requirements, with no recourse to considering how to make a universe hospitable to life.

Despite his advocacy for anthropic reasoning, Weinberg shared that sentiment. "The ideal would be to find that all of the constants of nature are uniquely fixed by some very small set of fundamental principles like the basic principles of quantum mechanics, and that there's no freedom at all, that every one of them can be and will be predicted," Weinberg said when I interviewed him in 1989. "That would be the ideal that we hope for. But maybe that's not true. It may be that there's some constants that can't be predicted in that way and for which we have to find an anthropic explanation."[21]

In his talk in Cleveland, Weinberg noted that part of the scientific enterprise was determining just what aspects of reality could in fact be explained from basic principles. In the early days of modern science, for instance, Kepler tried to deduce the Earth-sun distance from a basic theory—in his case, a theory based on three-dimensional geometrical objects. But no such theory dictates star-planet distances. If a star hosts a bunch of planets, one of them is likely to occupy an orbit making it habitable. There's no mystery to explain. "That's part of the progress of science—we learn what we can explain and what we can't," Weinberg said.

He emphasized, however, that it was premature to give up hope of finding a basic theory that would specify why the vacuum energy is as dense as it is. And many experts believed that the best hope for such a theory had already been identified.

Superstring theory, with roots extending back to the 1960s, had burst into prominence in the mid-1980s as the best candidate yet for a theory of everything. Many theorists believed that superstring theory would vaccinate them against anthropic infection. Superstring theory's main attraction was its ability to seamlessly merge gravity (Einstein's general relativity) into the same mathematical structure with quantum mechanics. Whereas standard quantum math conceived nature's basic subatomic particles as mathematical points, string theory regarded them as little snippets of string—one-dimensional objects, rather than zero dimensional points. Nature's whole zoo of tiny particles could be generated from different vibration modes of fundamental strings, kind of like different musical notes can be produced by a violin string. String theory seemed to possess the power—if anybody could work out some very complicated math—to fill in all the blanks in the mathematical description of the cosmos. Many physicists surmised that string theory could snuff out the anthropic principle by providing all the numbers needed for the complete cosmic blueprint, including how much vacuum energy there should be.

Except that it didn't. Even worse, string theory turned out to do just the opposite. Its math did not specify a single cosmos but described countless many. String theory more or less demanded a multiverse.

EXPLORING THE LANDSCAPE

While several investigators, including Linde, had suspected something along these lines earlier, the landmark paper implying a superstring multiverse appeared in 2000. In that paper Joe

Polchinski and Raphael Bousso showed that superstring theory allows innumerable different vacuum states—realms of space differing from one another in a vast range of properties, such as the number of dimensions, the strength of basic forces and, of course, densities of vacuum energy. In this "landscape" of possible vacuums, vacuum energy density could range across all possible values. And that would explain why no theory had been able to predict the "right" amount of vacuum energy—there was no single right answer. Vacuum energy's density may be an environmental accident—differing from place to place in the landscape—not an amount specified by basic physical law. In that case physicists could ask questions about vacuum energy's magnitude only in a space where that vacuum energy, presumably Einstein's cosmological constant, was in the right range to allow life to exist.

Polchinski, who sadly died in 2018 at the age of 63, was a leading superstring theorist at the Kavli Institute for Theoretical Physics at the University of California, Santa Barbara. He had been committed to the program of finding the ultimate equations describing nature and specifying things like the value of the cosmological constant.

"People in string theory were very fixed on the idea that there was some powerful mathematical structure we hadn't fully identified, and when we did, we would know why the cosmological constant was exactly zero," he recalled when I interviewed him in 2006. Even when cosmological acceleration was reported in 1998, "very few string theorists either knew or wanted to admit the significance of it in terms of the anthropic principle," he said. Nevertheless Polchinski began exploring the question in collaboration with Bousso, now at the University of California, Berkeley. Their math produced the inescapable conclusion that string theory specified not one but many possible values for the cosmological constant. While Weinberg showed that life's existence implied a measurable magnitude for the cosmological constant, neither he nor anybody else had identified the physics that could produce the right amount. Polchinski and Bousso's paper provided the physical model that could give you whatever magnitude of cosmological constant you wanted.

But Polchinski was dismayed. He had once said that if the cosmological constant turned out to be greater than zero, he would give up physics, because that would imply the need to embrace the anthropic dark side. But Bousso insisted that the paper needed to be shared with the world. After it appeared, in the *Journal of High Energy Physics,* others began to investigate what was soon referred to as the "string landscape": a vast array of vacuum-state possibilities. Among the scientists most intrigued was Leonard Susskind of Stanford.

"Lenny came along and said, 'Look, we can't sweep this under the rug; we have to take this seriously,'" Polchinski said. "'If this is the way things are, science is only going to move forward by thinking about it, not by pretending it's not there.'"[22]

Bousso and Polchinski's huge number of vacuum states emerged from the many convoluted ways that string theory's components could twist themselves up to make space. Superstring math requires more dimensions of space than the three commonly experienced in daily life—perhaps six or seven "extra dimensions" beyond the usual height, width, and depth. Presumably those extra dimensions curled themselves up into spaces too small for even an atom to notice. And they could curl themselves up in a huge number of different ways, often leaving "gaps" in space, kind of like doughnut holes, that Polchinski referred to as "handles."

Besides all that, one-dimensional strings might not be the only objects living in those dimensions. Two-dimensional "membranes" (think of the surface of a soap bubble) could vibrate within the higher dimensions, as might other objects—generically known as branes—with even more dimensions. Add to that the string theory equivalent of magnetic fields, known as fluxes, and a string universe could take on an impressive number of different identities. Any given arrangement of strings, branes, fluxes, and handles will determine the physical properties of the resulting space. And just as only three particles—neutrons, protons, and electrons—can combine to form more than a hundred different atoms (and millions of molecules), branes, fluxes, and handles can configure themselves into a megazoo of spacetime species, typically estimated to number something like 10^{500}.

Of course, it was possible that the superstring math was merely math, without any relationship to physical reality. But in 2003 Linde, with collaborators Shamit Kachru, Renata Kallosh, and Sandip Trivedi, analyzed the landscape in a paper published in *Physical Review D*.[23] They showed that many of the landscape's vacuum states very well might exist and could last much longer than the age of the known universe, allowing plenty of time to provide habitats for life.

It was immediately obvious to physicists working in this arena that the string landscape's many possible vacuums could correspond to the multiple bubbles produced by inflation. Linde had speculated almost exactly that in his very first paper on eternal inflation back in 1986. So string theory simultaneously bolstered both the validity of anthropic reasoning and belief in the reality of the multiverse. (On the other hand, in a 2006 paper Stephen Hawking and his collaborator Thomas Hertog suggested a connection between the string landscape and the alternative realities in the Many Worlds Interpretation of quantum mechanics.[24])

ANTIANTHROPIC ANIMOSITY

Support for the multiverse from string theory did not convert the anthropic opposition. At a physics conference in Newport Beach, California, in 2006, the philosophical schism in physics was on full display.[25]

Earlier that year, Nobel laureate Burton Richter had fired a shot at anthropic advocates in a letter to the *New York Times* book review. He excoriated the anthropic approach, describing its proponents as physicists who had "given up" on doing real physics.

"I can't understand why they don't take up something else—macramé, for example," Richter wrote. On a panel at the Newport Beach conference, Richter was just as acerbic.[26]

"The anthropic principle is an observation, not an explanation," he said. "The landscape, as far as I can see, is pretty empty. . . . It looks to me that much of what passes for theory these days is more like theological speculation." He didn't mind some physicists wasting their time on it, he said. "I don't see any problem with part of the theory community going off into a metaphysical wonderland, but I worry that it may be leading too many of the young theorists into the same thing." He lamented that the products of the anthropic approach had "no testable consequences."

Frank Wilczek, also a Nobel laureate, took issue with Richter's accusations. "I agree with the sentiment that Professor Richter expressed, that if you're not making statements about the physical world that can be tested in some way, shape or form, then you're

Andrei Linde (left) defends anthropic reasoning in cosmology during a panel discussion in 2006 as Burton Richter (center) and Leonard Susskind (right) listen.

not doing justice to what physics ought to be," Wilczek said. "But I don't accept the indictment. All of us have been trying to make predictions. . . . We're working very hard to try to turn this lemon into lemonade."

Linde responded even more vigorously in defense of anthropic reasoning. "It's science," he declared. "It's not science fiction. It's not religion. . . . It's something where we can really use our knowledge of mathematics and physics and cosmology." And he rejected the notion that anthropic theorizing is "giving up" and taking the easy way out. "It's complicated," Linde said. "It's not an easy job to do, so if you don't want to do it, then don't do it. But don't say that it's not science."

Making predictions is certainly an important part of physics, Linde agreed. But he emphasized that science is not just about predicting, but also about explaining. "You should not neglect one because you concentrate on another."

And not all explanations are alike. The total volume of interstellar space, for instance, is much greater than the volume of all the planets in the universe, Linde pointed out. Yet we live on a planet. That's not a prediction from some fundamental law; we live on a planet instead of in interstellar space because there's no oxygen to breathe in interstellar space. Environmental conditions explain where we live. "If we have an environmental explanation of where we are right now, then . . . it does not force us to continue studying why we are not in interstellar space," Linde said. "It's enough to know that we can live in the place where oxygen is. So there are some explanations of properties by the anthropic principle and they cannot be neglected."

Richter responded by saying life's presence here is an observation, not an explanation. Whereupon Leonard Susskind interrupted: "Burt, you don't seem to understand what everybody has said here." Richter remained unpersuaded. As in centuries past, agreement among experts on the nature of the universe remained elusive.

During the conference, I interviewed Susskind in more depth about the anthropic debate. In his view, though, there was no debate—just emotion-laden responses to some recent purely scientific developments. "There's no substantive scientific debate," he said. "What's going on is different emotional reactions to some facts and some interpretations of those facts that we've discovered." Those facts had made it clear to nearly everybody that physics has to accept some environmental explanations. "Some features of physics may be environmental," Susskind said, "and we have to figure out which ones."[27]

The many "environments" in the cosmos implied by the string landscape were not invented to support anthropic reasoning, Susskind pointed out. They "materialized out of the mathematics." It's a key point, reminiscent of arguments posed by the Greek atomists favoring the existence of innumerable worlds. String theory supposedly describes reality's tiniest building blocks, the "strings" that vibrate as fundamental particles that make up atoms, the components of the material world. When the Greek scholars Leucippus and Democritus developed their analogous theory of atoms and void (as discussed in chapter 3), they realized that the theory itself logically required the construction of multiple universes (or *kosmoi*). To make the world as it appeared from the random jumblings of atoms, you needed an infinite number of them, interacting in a vast void. Once the world had formed, an infinite number of additional atoms remained, and they would inevitably create even more worlds. The conclusion of an innumerable number of worlds emerged from the theory; it was not simply an additional idea that the atomists happened to like. In a similar way, an innumerable number of vacuum states emerged from the mathematics of string theory; a multiverse of some sort is implied by the best theory of the subatomic parts that make up the world's materials. The string landscape came from the theory, not from the physicists studying it. As Polchinski told me, "String theory has over the years given us what it wants to give us, not what we expected it to."[28]

It was the emergence of the landscape that converted Susskind from an interested bystander to an advocate for anthropic cosmology and the inflationary multiverse. The landscape implied a mechanism for creating diversity in the cosmos. And inflation's success at describing the universe implied that it was much, much bigger than the part we can observe. That meant that there could be plenty of room for cosmic diversity to have so far escaped notice. It's an open question whether the rest of the universe is just like the part we see.

"We no longer have any evidence that our little piece of the universe is representative of the whole thing," Susskind said. "So then we have a question: Is the universe very, very big, with all the same rules everywhere . . . or is it a patchwork?" Either possibility is very difficult to confirm—or falsify. "For me, my own opinion is that anybody who has thought hard about it is leaning more toward the diversity than the complete homogeneity," Susskind said. "From all we know now, it is less likely that the universe is everywhere the same. Once we agree that it's diverse, then some features of it are environmental and some features of it will be determined by whether we can live in it or not."[29]

It's possible, of course, that features initially appearing to be environmental might someday receive a more fundamental explanation. David Gross pointed out to me that energy levels in atomic nuclei had once seemed inexplicably random.[30] But eventually the theory of the strong nuclear force showed how those nuclear properties were specified by fundamental laws. Besides, Gross noted, both the theories of inflation and the superstring landscape remain works in progress. Neither have been established at a sufficient level of rigor that scientists should blindly accept their consequences.

Yet it seemed to many others it would be equally foolish to ignore the possibility of those consequences. After all, the prospect of universes beyond the one now known was not unprecedented. "Again and again in the history of cosmology, we've been shown that the little pieces we've been looking at are not the whole story," cosmologist Sean Carroll of Caltech pointed out. "There's a boundary to what we can see, and we should be open-minded to what exists beyond that boundary."[31]

Today it remains uncertain whether other universes exist beyond the reach of our sight. But suppose for a moment that they do. Countless bubbles, containing every possible permutation of physical properties, populating the entire expanse of space and time. Surely, then, the story comes to an end, with no more prospect of anything beyond. No more universes to conquer. Right?

Guess again.

Brane Worlds

> We were able to realize that this universe in which we lived not
> only extended but was, as it were, slightly bent and contorted,
> into a number of other long unsuspected spatial dimensions. It
> extended beyond its three chief spatial dimensions into these
> others just as a thin sheet of paper, which is practically two
> dimensional, extended not only by virtue of its thickness but
> also of its crinkles and curvature into a third dimension.
>
> —H. G. WELLS, *MEN LIKE GODS,* 1923

N ONE OF HIS LESSER-KNOWN NOVELS, H. G. Wells
wrote of an adventure in the life of a journalist named Alfred
Barnstaple. While on a vacation drive through England one day
in the fall of 1921, Mr. Barnstaple suddenly found himself in a new
world. Scientists in a parallel universe had accidentally plucked him
and a few other humans from Earth and deposited them in a socially and technologically advanced civilization known as Utopia.

A Utopian scientist named Serpentine attempted to explain
the situation to the Earthlings. Utopians had discovered that
their supposedly three-dimensional universe occupied a higher-
dimensional space. They concluded that other universes could
exist in that space, too. "Just as it would be possible for any number
of practically two-dimensional universes to lie side by side,
like sheets of paper, in a three-dimensional space," Serpentine
said, "it is possible for an innumerable quantity of practically

three-dimensional universes to lie, as it were, side by side and to undergo a roughly parallel movement through time."

One of the humans, Mr. Burleigh, responded for the group with a limited degree of understanding. "We accept your main proposition unreservedly," Mr. Burleigh said, "namely, that we conceive ourselves to be living in a parallel universe to yours, on a planet the very brother of your own." And that sentence, in Wells's *Men Like Gods,* seems to be the first use in science fiction of the phrase "parallel universe."[1]

It was not, obviously, the first time any writer of fiction or fact had imagined other universes. But this other universe was a different sort. Not another set of spheres beyond the firmament. Not a system of planets around a very distant star. Not an island of stars far enough away so as to appear a fuzzball through powerful telescopes. Not even a huge expanding bubble of spacetime inflated from a distant part of the same space.

No. Wells's new universe was literally parallel to ours. It existed not far away, but nearby—in another dimension. Mr. Barnstaple, Mr. Burleigh, and the other humans who went there did not fully grasp its nature. But modern physicists know exactly what Wells was describing. Today they would call it a "brane world."

OTHER DIMENSIONS

While Wells may have been the first to use the term "parallel universe," he was not the first to contemplate the existence of other dimensions. Way back in 1747, a young Immanuel Kant discussed the possibility of more than three dimensions of space. He argued that Newton's law of gravity, with the strength of gravitational attraction diminishing with the square of the distance between two masses, would be correct only in a space of three dimensions.[2] Kant acknowledged that other dimensions might be possible in principle; if so, he surmised, then God would no doubt have "somewhere brought them into being." But they would not be part of our universe. "Spaces of this kind, however, cannot stand in connection with those of a quite different constitution," Kant wrote. "Accordingly such spaces would not belong to our world, but must form separate worlds."

That sounds very much like the modern idea of parallel universes. And Kant was explicit that he did not mean separate worlds metaphorically, but an actual multiverse. "In the considerations now before us [of extra spatial dimensions] we have, as it seems to me, the sole condition under which it would also be probable that a plurality of worlds actually exist."[3]

About a century later, the German psychologist Gustav Fechner (1802–1887) wrote a short story illustrating the idea that space could have four dimensions. More famously, the English schoolteacher Edwin Abbott Abbott explored similar ideas in 1884 in his clever novel *Flatland*.[4] Abbott set his story in a world with only two dimensions (hence Flatland), whose male inhabitants were geometrical figures like triangles and squares. Women, though, were permitted only one dimension, existing as lines (it was Abbott's way of satirizing Victorian-era discrimination against women).

Flatland's world illustrated the limits of thinking only within the box of the dimensions you can see. One day a Flatlander mathematician named A. Square noticed a circle growing larger and then smaller again. He eventually figured out that he was witnessing the intersection of a sphere from a higher dimension with his own two-dimensional universe. As the sphere rose and fell, the circle marking its points of intersection with Flatland grew and shrank, being biggest when the sphere's equator coincided with Flatland. A. Square was able to deduce from his observations (with some help from the sphere) that Flatland was not the whole universe; its two-dimensional plane existed within a grander three-dimensional space. When he tried to share his momentous discovery with other Flatlanders, they threw him in jail.

Flatland was, of course, a fictional fantasy. But the idea of an extra dimension had real-world implications, as the American astronomer Simon Newcomb noted in 1893. In an address delivered to the New York Mathematical Society, he discussed the possibility of a multiverse in a geometry with four dimensions of space rather than three.

"It is, therefore, a perfectly legitimate exercise of thought to imagine what would result if we should not stop at three dimensions in geometry, but construct one for space having four," he said. "Add a fourth dimension to space, and there is room for an indefinite number of universes, all alongside of each other, as there is for an indefinite number of sheets of paper when we pile them upon each other."[5]

It's easy enough to imagine a set of parallel sheets of paper. Just look at a book. Every page extends a sensible distance in two dimensions (top-to-bottom and left-to-right)—a two-dimensional world, inhabited by alphabetical letters and punctuation marks. Each page is a parallel world, very nearby but distinct from the pages before and after. These book-page universes are two-dimensional from the ink's point of view. But each sheet does in fact possess a third dimension, just a very thin one. So hundreds of pages can fit into a space much smaller than the two dimensions of the pages themselves. As Newcomb realized, something similar might be the case with the three-dimensional space that humans are aware of. Space—the universe—might also be concealing from us a very tiny fourth spatial dimension. Countless universes like ours could conceivably exist, parallel to each other, in the additional dimension.

But does such an additional dimension actually exist? Newcomb couldn't say. "From the point of view of physical science, the question whether the actuality of a fourth dimension can be considered admissible is a very interesting one," he said. "All we can say is that, so far as observation goes, all legitimate conclusions seem to be against it." Nobody had seen any sign of such a dimension, or observed any phenomenon that suggested it, he pointed out. "If there is another universe, or a great number of other universes, outside of our own, we can only say that we have no evidence of their exerting any action upon our own."[6]

A HOME FOR A MULTIVERSE

Both Abbott and Newcomb were probably influenced in their views by a writer who took the idea of another dimension more seriously than anybody else. His name was Charles Howard Hinton. He was a curious character, trained in math, who wandered the world leaving a wake of various personal problems (such as a conviction for bigamy) while obsessing over the fourth dimension.

Hinton (1853–1907) popularized his ideas in a series of what he called "scientific romances." First came "What is the Fourth Dimension?" in 1884, the same year as Abbott's publication of *Flatland*. Nowadays, of course, the fourth dimension refers to time, as required by Einstein's theory of relativity. But for Hinton, the fourth dimension meant another dimension of space. (For Wells, curiously, the fourth dimension did mean time, as he described it in his famous novel *The Time Machine,* published a decade before Einstein's 1905 relativity paper and almost three decades before *Men Like Gods.*)

Hinton believed that extradimensional reality could manifest itself in more than just a growing and shrinking circle. All sorts of phenomena observed in the familiar three dimensions of space, he explained, might just be the visible effects of objects operating in an additional dimension.

Perhaps "the movements and changes of material objects" are produced as our three-dimensional world intersects a "a four-dimensional existence." Imagine, Hinton said, pulling a thread perpendicularly through a thin sheet of wax. Then grasp the thread, from above and below the plane of the wax, and pull it along the sheet at an angle. "If now the sheet of wax were to have the faculty of closing up behind the thread, what would appear in the sheet would be a moving hole." The speed of the moving hole would depend on the angle of the thread.[7]

Two-dimensional physicists living in this wax paper flatland would theorize that the moving hole was a particle of some sort.

Without awareness of the higher dimension, the flat physicists would know nothing about the thread.

Of course, there is no reason to stop with one thread. So Hinton would next have you imagine, more abstractly, many straight lines from higher dimensions extending through any plane, at various angles. As those lines moved, "there would then be the appearance of a multitude of moving points in the plane, equal in number to the number of straight lines in the system." If you think of those points as atoms, you can see how the movement of a plane through a system in a higher dimension could create a rather complicated reality. In the same way, perhaps our three-dimensional universe acquires its complicated physics from its intersection with higher-dimensional objects, Hinton speculated. In other words, a permanent four-dimensional system passing through our 3-D space could be responsible for the ephemeral motions and changes we witness in the material world. "Change and movement seem as if they were all that existed. But the appearance of them would be due merely to the momentary passing through our consciousness of ever existing realities," Hinton declared.[8]

Hinton realized that our 3-D space might not be the only one that could intersect with a higher dimension. Even if our universe happened to be infinitely large (in three dimensions), it could still fit in the fourth dimension with space to spare. It would be like a book with infinitely wide pages—they still could be stacked in the extra dimension defined by the book's spine. So an infinite-in-size three-dimensional object—coextensive with the entire universe—would not fill all of a space possessing a fourth dimension. "There could be in four-dimensional space an infinite number of such solids, just as in three-dimensional space there could be an infinite number of infinite planes," Hinton pointed out.[9] Conceiving the infinite three-dimensional object as a universe, this seems very much like saying the existence of an extra dimension provides a home for a multiverse.

But like Newcomb, Hinton was not willing to declare with certainty that such a dimension existed. "No definite answer can be returned to this question," he wrote. He reasoned that it might be possible to find out, though. If we exist in a fourth dimension, our extension in it must be extremely minute, or we would be conscious of it. But the extra dimension might reveal itself to "the ultimate particles of matter," which would be minute enough for the fourth dimension to seem comparable in size to the other three.[10] And that really sounds kind of like what some modern physicists say about the situation in physics now. If Hinton were alive today, he'd be a string theorist.

THE FIFTH DIMENSION

Stories by Hinton (and Abbott, and Fechner, and others) illustrated the idea of extra dimensions and explored their ramifications generally, but not technically. In a more serious mathematical sense, extra dimensions had already emerged from the mind of Bernhard Riemann (1826–1866) in 1854. Riemann was interested in the foundations of geometry. He developed a new version of geometry, differing from Euclid's, incorporating the tools for describing multiple-dimensional spaces, called manifolds. Riemann did not discuss the likelihood that real space contained any of these extra dimensions, though, let alone whether any additional universes might populate them. A few other scientists, such as William King Clifford and Hermann von Helmholtz, also wondered about the math and science of extra dimensions.[11] But as far as I can tell, Newcomb was the first scientist to specifically mention higher dimensions as a repository for multiple universes. (That's excluding the "scientists" of late nineteenth-century England who speculated that extra dimensions were spirit worlds responsible for ghosts and other paranormal phenomena. They don't count.)

As it turned out, extra dimensions had other scientific uses besides providing a home for new universes. In 1912, for instance, the Finnish physicist Gunnar Nordström (1881–1923) proposed that a fourth spatial dimension might be useful for a theory explaining both gravity and electromagnetism. Soon thereafter, though, Einstein explained gravity without any need for extra space dimensions, and Nordström's theory faded away into some other unknown dimension. Then a few years later, Theodor Kaluza (1885–1954) revived the idea of an extra dimension by adding one to Einstein's gravity, suggesting it was just what the doctor ordered (I presume it was Doctor Who) to explain electromagnetism.

Kaluza, working at the University of Königsberg, rewrote Einstein's equations for general relativity in five dimensions instead of four. With the extra dimension, the gravity equations could be reformulated into Maxwell's equations for electromagnetism. In 1919 Kaluza sent his paper to Einstein, who arranged to have it published.

It's not clear that Kaluza regarded his fifth dimension as real rather than just a mathematical trick. In any event, there was no evidence that such a dimension actually existed—it had no detectable influence on any known physical process, and nobody named Barnstaple had ever visited it.

Meanwhile, though, the Swedish physicist Oskar Klein was off visiting the University of Michigan, where he was also contemplating a fifth dimension. When Klein returned to Europe in 1926, he learned of Kaluza's work and despaired at his idea having been anticipated. But Klein did take the idea of the fifth dimension seriously as an actual new dimension of real space. And he had a good idea why nobody ever noticed Kaluza's extra dimension. It was too small to see.

Normally, in ordinary space, you never think about a dimension having a particular size. Space seems limitless no matter which direction you fly—every dimension is immensely big. If you travel in an ordinary dimension in a given direction, you just keep going. But if a dimension had curled up on itself to become small, you would eventually get back to where you started from. Think about traveling on the Earth's surface, where the dimensions are designated by longitude and latitude. As you travel around the Earth at the equator, maintaining the same latitude but moving through different longitudes, your trip will take a very long time—at the equator, longitude and latitude are big dimensions. But now take a trip around the world a few inches from the North Pole. You can complete your trip in an instant, because near the pole the dimension you are moving through is very small. You can see that as you get closer to the pole, that dimension gets smaller and smaller, eventually too small for anyone to possibly notice.

Klein envisioned just such an almost inconceivably small dimension in addition to the three big dimensions of space. It was curled up, into a tiny circle, far too small for the eye (or any microscope) to see. He calculated a circumference of about 10^{-30} centimeters—smaller than an atom to the same extent that an atom is smaller than the solar system. That would explain why nobody had found evidence that an extra dimension actually existed. Of course, Klein's math didn't provide any evidence that such a dimension did exist, either.

So Kaluza and Klein's idea did not exactly inspire extra-dimension mania. But it did attract some interest; at one point Einstein even wrote a paper on it. It was also taken seriously by the famous Austrian physicist Wolfgang Pauli. But nothing much came of those efforts. Kaluza-Klein theory was never completely forgotten, though. And a few decades later, it came in handy when physicists began struggling with superstrings.

STRING FEVER

Initially, in the 1960s, string theory was an idea for explaining the strong force that held particles together in an atomic nucleus. It didn't quite work. Then in the 1970s another approach, called quantum chromodynamics, succeeded nicely in describing the strong nuclear force, so the string strategy was no longer needed anyway. But while string theory failed to explain nuclear attraction, it exerted a strange psychological attraction for some physicists and never really died. A parallel development of a concept called supersymmetry—conventionally referred to as SUSY for short— soon revived it. Supersymmetry put the "super" in superstring.

SUSY appealed to theorists' entrancement with the relationship between mathematical symmetries and the laws of nature. Einstein's relativity theory had established the power of symmetry to reveal physical insight into the cosmos. And then mathematical representations of various sorts of symmetries (a branch of math called group theory) had almost magically captured the physics of nature's fundamental forces and particles. Quarks and leptons (the building blocks of matter) and photons, Z and W bosons and gluons (the particles transmitting fundamental forces) all emerged naturally from the equations for various symmetry groups. The set of equations describing those symmetries became known in the 1970s as the Standard Model of particle physics.

During that time, some physicists began to wonder whether a new level of symmetry might describe a relationship between the matter particles and force particles. If so, this "supersymmetry" would imply the existence of an unsuspected new population of subatomic particles in the universe—one new force particle corresponding to each matter particle, and one new matter-particle partner for each force particle. That prospect led to a merger of the old string theory ideas with the supersymmetry concept, launching the saga of superstring theory.

Superstrings were conceived to be one-dimensional, like lines (hence strings), but extremely small, on the order of Kaluza's extra dimension. They seemed to be a little like rubber bands, vibrating within their higher dimensional space. Some of the strings were viewed as loops, but others were like rubber bands that had been snipped, leaving two loose ends.

A major boost for superstring theory came in 1984, when John Schwarz and Michael Green showed that superstrings removed some of the mathematical obstacles to creating a quantum theory of gravity. In fact, superstring theory actually predicted the existence of a particle with a spin of 2 (in subatomic spin units). That was just the sort of particle that would represent a quantum of the gravitational field—a particle responsible for transmitting gravitational force. Suddenly superstring theory became the hottest topic in particle physics; physicists proclaimed a "Superstring Revolution." Enthusiasm for superstring theory was tempered, though, by an intriguing mathematical complication: its math required space to possess many more than the known three dimensions.

Kaluza and Klein's extra "fifth" dimension was not nearly enough for superstrings. Depending on how the superstring approach was formulated, the math called for at least 10 dimensions, one of time, nine of space. At first some physicists suspected that those dimensions did not really need to exist—they merely served a mathematical purpose, like the ancient Greek spheres that predicted the motions of the planets. The spheres (or the strings) didn't need to be real for the math describing them to be helpful (you'll recall the notion of saving the phenomena). When I first wrote about superstrings in 1985, I posed that issue to Schwarz, suggesting that the strings and extra dimensions were like that— mathematical conveniences, not actually physical. "Oh no!" he exclaimed in reply. "They're intended to be real." It soon became clear to others that tiny strings and their extra dimensions did need to exist for the theory to represent reality. But everybody presumed that those extra dimensions were all very small, as Klein had calculated. Space's extra dimensions must have curled up, or compactified, into space that to our eyes and instruments disguised itself as merely three-dimensional.

SUPERMEMBRANES

As string theory developed, several physicists, among them Joe Polchinski, then at the University of Texas at Austin, realized that the math permitted more than just strings. With multiple dimensions to play around in, you can have objects possessing more dimensions, too. In other words, strings might not be the only characters on the multidimensional stage. Instead of just one-dimensional "rubber bands," you might have two-dimensional "soap bubbles." Conceiving the 2-D surface of the soap bubble as a membrane, it was natural to call such objects super-membranes.

Physicist Michael Duff, then in England but soon to move to Texas A&M University, liked that idea. He argued that instead of just strings in 10 spacetime dimensions, the theory should incorporate 2-D supermembranes in 11 dimensions. And supermembranes didn't need to limit themselves to two dimensions. You could have three-dimensional or five-dimensional or higher-dimensional inhabitants of the extra dimensions. All became known generically as branes; a two-dimensional supermembrane, for instance, was called a two-brane.

Some physicists didn't immediately grasp the beauty of the idea. When I asked Duff in 1990 about the reception to supermembranes, he replied that most string theorists preferred to stick to strings. "Hard-nosed string theorists would scoff at membranes, because they're dedicated to the string and that's the end of the story," Duff said.[12] String dogma settled on one-dimensional strings in 10 spacetime dimensions. Supermembranes in 11 dimensions seemed to require one dimension too many.

But within a few years, string dogma was shattered in what came to be known as the "Second Superstring Revolution." At a meeting in California in 2005, Edward Witten, one of the leading pioneers of string theory, took it to the next level—or more specifically, the next dimension. By thinking in 11 dimensions rather than 10, Witten resolved a longstanding puzzle that had perplexed superstring theorists. It was supposed to be the ultimate theory, the only theory needed to describe all of fundamental physics. But there were five distinct mathematical versions of superstring theory. All five yielded a consistent picture. It seemed obvious that five different theories couldn't all be the right theory. So maybe none of them were.

Witten, however, realized that in 11 dimensions, all five string theories could be shown to be equivalent. Superstring's five theories were just the same actor in different disguises, like Tom Hanks in *The Polar Express* (a film my sister insists on everybody in the family watching every Christmas). As Duff explained it to me, it was a little like the old story of the five blind zoologists and the elephant. Each theory represented a different part of the elephant's anatomy. Physicists had simply been blind in the 11th dimension. By going there, Witten could see the whole elephant, the umbrella theory that encompassed the other five. And adding that dimension brought supermembranes back into the story, seeding a new embryonic theory. Witten called it M theory, suggesting that the M could stand for magic, mystery, or membrane.

Almost immediately, membranes became much more useful than they had been. A key development came quickly from Polchinski (who had by then moved to the University of California, Santa Barbara). He had long been studying a particular type of brane, known as a D-brane (named for Peter Gustav Lejeune Dirichlet, a nineteenth-century mathematician). After Witten introduced M theory, Polchinski put the D-brane picture together in a new way with an astounding implication: strings lived on branes.

Stating it so simply conceals an immense amount of complicated mathematical analysis. But it's the essential idea. Snippets of superstring, with two loose ends, cannot function untethered. Each end has to be attached to something, and D-branes provide a suitable surface. That means that matter and the particles that transmit forces are anchored to branes. Gravity, on the other hand, is transmitted by loops of string without loose ends. Gravity can wander from brane to brane.

With these new developments, a novel chapter in the story of multiple universes began. If there's ever a movie version, it will be called "Braneworld."

BRANE WORLD COSMOLOGY

In 1996 Witten, collaborating with Petr Hořava, explored the 11th dimension (mathematically) and encountered a curious possibility. They proposed that an 11th dimension could serve as a space separating two 10-dimensional spacetime branes (each possessing nine dimensions of space, one of time). Each brane would be like a wall (yes, a nine-dimensional wall). The particles and forces known to physics might all live on one of those walls. The other wall would be, well, a parallel wall, a "hidden world." Basically another universe. Or a "brane world."

Our world does not appear to be nine-dimensional, of course. But eventually the view emerged that our three-dimensional space could be construed as a 3-D supermembrane, a three-brane. It's big in three dimensions but small in others, and so could fit easily into an extra dimension. That extradimensional space was called the bulk. Particles of matter and radiation (including light) would all be stuck on our three-brane because those particles would consist of strings whose ends needed to be connected to a brane. Gravity, on the other hand, transmitted by loopy strings, could fly in the space between the walls unimpeded. After all, gravity is the geometry of space—where there's space there's gravity. So you can't confine gravity to a brane. It lives in the bulk. In this picture our universe is a three-dimensional bubble, or brane, floating in the bulk.

But wait a minute. The curled up extra dimensions of superstring theory are supposed to be very small. How could a universe-sized three-brane fit in? Well, possibly, it would be just like the way a two-dimensional sheet of paper could fit inside a very thin book. Or the three-dimensional objects that Hinton imagined to be infinitely big. They would be thin in another dimension.

But it's also possible that string theory's extra dimensions aren't all supersmall.

Physicists had mostly believed that Klein must have been right about the tiny size of an extra dimension. If an extra dimension was relatively big, laws describing phenomena such as electromagnetism would no longer work. Gravity would also behave badly if other big dimensions existed. Newton's inverse square law (retained by Einstein's general relativity) depends on the presence of precisely three dimensions of space. (Immanuel Kant had made that point 250 years earlier.)

But often in physics, loopholes can be found in such no-go arguments. Savas Dimopoulos of Stanford and colleagues found one of those loopholes. From some previous studies, Dimopoulos had learned that the strength of gravity had never been precisely tested on very small scales. It was possible that gravity did not obey the inverse square law at distances smaller than a millimeter or so. Dimopoulos and colleagues Nima Arkani-Hamed and Gia Dvali worked out the details of that idea in a paper published in 1998.[13]

I attended a conference the next year at Fermilab, where Dimopoulos presented the case for a (relatively) large extra dimension, perhaps as big as a millimeter, as a bulk for branes to float in. He explained that many of the problems facing physics at that time—such as the small mass of the neutrino and the weakness of gravity compared with other forces—could be solved by the presence of a big extra dimension. "A large number of these theoretical questions can be answered by physics that actually takes place in the bulk," he said.

And there was more. Our bubble might not be the only one. "It might be that we are not alone in the bulk—maybe there are other three-branes that populate the bulk," he said. "And each one of them are three-dimensional spaces as good as ours."[14] Here was a new kind of multiverse—parallel cosmic bubbles, merely a millimeter away, in another dimension.

It was especially attractive that this idea could be tested by searching for deviations from ordinary gravity at submillimeter distances. And in fact, that search has been conducted, although so far with no sign of any effect on gravity, even at length scales much shorter than a millimeter. But that doesn't rule out extra dimensions or brane worlds.

At the 1999 Fermilab conference, the next speaker after Dimopoulos was Lisa Randall, then at MIT and now at Harvard. I had been aware of Dimopoulos's work on millimeter-sized extra dimensions and followed it closely. But Randall's talk took me by surprise. She too had ventured into the realm of another dimension, and returned to report that it might be even bigger than a millimeter. A lot bigger. "We have either small extra dimensions," she said, "or an infinite extra dimension."[15]

Randall, collaborating with Raman Sundrum, showed that gravity might prefer to concentrate on a brane other than ours. In that case gravity would be relatively weak on our brane (and it is) but would not be altered in strength by even a very large extra dimension, perhaps infinitely large. We would still notice only our usual four-dimensional world, three of space and one of time.

"I'm going to argue," Randall said at the conference, "that we can actually have an infinite extra dimension but a theory that really looks four-dimensional." In fact, the world she described would

Lisa Randall

appear to be so precisely four-dimensional that the extra dimensions could elude detection by testing for short-distance deviations in gravity. While a millimeter-sized dimension could reveal itself in that way, Randall and Sundrum's infinite dimension would not. "When you have small extra dimensions you have experimental consequences," Randall explained, "but when you have an infinite extra dimension you have no experimental consequences."[16]

As Randall and others developed the technical details of this idea further, they found a variety of brane world scenarios. A few

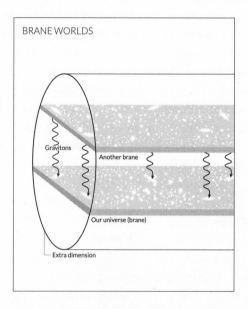

BRANE WORLDS

Gravitons

Another brane

Our universe (brane)

Extra dimension

Brane Worlds. If extra dimensions of space exist beyond the three commonly known, then our universe may be just one of many within the extra-dimensional space. Objects residing within such a space are known as branes (short for membrane). Our three-dimensional space—a three-brane—could float in an extra dimension parallel to another similar universe. Ordinary radiation and matter would be confined to the branes, but gravity—transmitted by particles called gravitons—could travel from one brane to another.

years later, in a popular book, Randall wrote that those scenarios implied a revolutionary new view of the universe, comparable to the upheaval perpetrated by Copernicus.

"Not only are we not in the center of the universe, as Copernicus shocked the world by suggesting five hundred years ago, but we just might be living in an isolated neighborhood with three spatial dimensions that's part of a higher-dimensional cosmos," Randall wrote. And our three-dimensional neighborhood—our brane—might not be alone. "Other branes might be parallel to ours and might house parallel worlds," she speculated.[17]

Some of those parallel worlds might be three-branes, similar to our universe; others might be unimaginably different. "Braneworlds introduce new physical scenarios that might describe both the world we think we know and other worlds we don't know on other branes we don't know, separated from our world in unseen dimensions," Randall declared. "If there is life on another brane, those beings, imprisoned in an entirely different environment, most likely will experience entirely different forces that are detected by different senses."[18]

In that sense, the multiverse of the brane world could be much more bizarre than the plurality of worlds traditionally imagined.[19] Christiaan Huygens (chapter 8) believed life on other worlds would require similar senses to humans (not to mention hands and feet), as light, sound, and touch were such important ways of learning about reality. But in brane worlds, new kinds of geometry, new kinds of matter, new realities may await any future Mr. Barnstaples lucky enough to take a wrong turn into another dimension.

If extra dimensions and brane worlds truly exist (and Randall, for one, believes they do), then one more step gets added to the ladder of the hierarchy of multiverse scenarios. The inflationary bubbles, the infinite froth of cosmic carbonation, become merely the flatland polygons on a single three-brane, a spaceworld floating in an even vaster space. Once again, the word (and concept) of universe would need to be redefined.

Defining the Multiverse

I can imagine that the ultimate future of the physical sciences may
lie not so much in unraveling the properties of our particular
environment as in working through the possibilities permitted by all
the kinds of environments that may exist in a universe of far greater
complexity than we contemplate in our usual cosmological studies.

—FRED HOYLE, *AMERICAN SCIENTIST,* 1976

MULTIVERSE is a term with multiple meanings.

Its first use seems to have been by the philosopher
William James. In an address delivered to the Harvard YMCA in
1895, he spoke of the moral mixture that nature presents to
human existence. Our sacred books tell us that God made heaven
and earth and saw that they were good, James noted. But the realities of life "refuse to be brought by us into any intelligible unity
at all." Instead "beauty and hideousness, love and cruelty, life and
death keep house together in indissoluble partnership." Rather
than unity, reflecting the moral goodness of a god, nature offers a
moral diversity. "Visible nature," James proclaimed, "is all plas-
ticity and indifference, a moral multiverse, as one might call it,
and not a moral universe."[1] From James's moral multiverse stems
today's common dictionary definition of *multiverse:* "a totality of
things and forces that are disparate or lacking in ultimate unity."[2]

James, of course, used multiverse in a moral-philosophical
sense. He wasn't thinking about cosmology. A similar usage, but

closer to a scientific meaning, was expressed in 1904 by Oliver Lodge, the British physicist. Events in the cosmos might be the natural playing out of the laws of physics in "an orderly and systematic universe," Lodge said. Or we may occupy a spiritually guided universe, purposeful and directed, "carrying on its evolutionary processes . . . with a definite aim . . . permeated throughout by mind and intention and foresight and will." A third possibility was "a universe of random chance and capricious disorder, not a cosmos or universe at all—a multiverse rather."[3]

In a more clearly astronomical sense, *multiverse* was used as early as 1915 in reference to the possibility of multiple "island universes," the long-suspected remote and independent cousins to the Milky Way—each a separate universe as the notion of universe was then conceived. E. L. Eaton, chair of astronomy at the University of Wisconsin, used *multiverse* in this way. During a lecture he showed pictures of clusters of stars "which some have designated as Multiverse, in distinction from universe." Those clusters "seem to be altogether outside the Milky Way universe, and probably millions of other clusters are to be found by the larger instruments to be constructed for photographing the heavens."[4]

After Edwin Hubble confirmed the existence of island universes, Harlow Shapley used the term *multiverse* to refer to them in popular talks.[5] But *multiverse* never caught on as a general description of the new expanding universe containing billions of those island universes. The island universes became known as galaxies, all residing within a single universe once again redefined—a vast expanding bubble of spacetime. In the 1980s, when theories of eternal inflation implied the existence of many such bubbles, *multiverse* once again was called on to describe the new cosmological conception.[6]

AN ATOM IN THE GREATER REALITY

Curiously, the term *multiverse* itself—for describing something like the many universes of inflation theory—showed up in print before the theory even existed. The earliest example I can find is in a story by the legendary science fiction writer Jack Williamson, in the July 1978 issue of *Analog Science Fiction/Science Fact*. Titled "Kinsman to Lizards," the story describes a future of genetic engineering where "premen" have genetically created intellectually superior "trumen," who eventually realize that they all live in a multiverse. At one point a character named Davey asks somebody named Pipkin (apparently an immortal god) if he can explain the multiverse. "If our own universe goes on forever, what can be outside it? That's the sort of thing I need to understand," Davey says.

"Your preman forebears had a theory of a single universe," Pipkin answers. "An explosion of energy and mass, creating space and time as it swells to its gravitational limits, erasing them as it falls back into the point of its beginning, recreating itself as it explodes again—"

Is that not true? Davey interjects. Pipkin replies: "That theory is a fair enough fit for this one universe, which is all the premen and the trumen are able to sense, though it's only an atom in the greater reality. The actual multiverse holds an infinity of such universes, all held within a wider domain of order that the gods can perceive and mortals cannot."[7]

After inflationary cosmology came along a few years later, scientists gradually adopted *multiverse* to describe its many space-time bubbles, although occasionally you would see "megaverse" or "metaverse" instead. Eventually *multiverse* prevailed. And it began to show up in other contexts, such as a label for the many worlds of Hugh Everett's interpretation of quantum mechanics. Later, scientists adopted *multiverse* in connection with the superstring landscape and sometimes also for the parallel brane worlds that might occupy superstring theory's extra spatial dimensions.

And so today *multiverse* can mean very different things. Modern physicists do not contend with merely one multiverse; they deal with many possible versions. It's even possible that more than one of those various types of multiverse actually exist. Nature may turn out in the end not to be best described by a multiverse, but by a multiplicity of multiverses. (I'm tempted to coin a new word for that—perhaps *multiplexiverse*—but maybe that job should be left for science fiction writers.)

It's starting to sound complicated. But as the term *multiverse* has evolved in modern physics and cosmology, its meaning no longer really depends very much on what kind of multiverse you're talking about. What makes the multiverse worth wondering about is not so much its origin as whether it provides a diversity of conditions in different domains of reality. It's the diversity that makes the multiverse scientifically worthwhile. And that makes it (despite the complaints of multiverse deniers) scientific.

A MULTIPLICITY OF MULTIVERSES

Max Tegmark, you'll recall from chapter 12, classified different multiverse versions into levels. Level I refers simply to large regions of space beyond the observable universe's horizon. Level II refers to regions that vary in their local "effective" laws of physics, possibly corresponding to the spacetime bubbles produced by chaotic inflation. Level III corresponds to the many worlds produced in the branching quantum wave function of the universe, as described by Everett's interpretation. Tegmark also argued for a Level IV multiverse: the set of all consistent mathematical structures. He proposed that all such structures, in some sense, exist as physically real universes.[8]

An even more comprehensive classification of various multiverses appeared in physicist Brian Greene's book *The Hidden Reality*. He identified at least nine ways to conceive of a multiverse:

> Quilted multiverse: in distant regions of infinite space, configurations of matter repeat themselves, creating parallel worlds.

Inflationary multiverse: never-ending inflation produces a vast network of expanding spacetime bubbles; our universe is one of them.

Brane multiverse: our universe is a three-dimensional membrane (or brane) floating in another dimension, which may include additional three-branes parallel to ours.

Cyclic multiverse: collisions between branes can create new big bangs, generating a series of universes sequential in time.

Landscape multiverse: string theory (or M theory) implies multiple possible vacuum states arising from different compactifications of extra dimensions; those different vacuums might be realized in inflation's many spacetime bubbles.

Quantum multiverse: many branches of different possibilities contained in the quantum wave function describing reality all exist in some sense, creating branching worlds or universes.

Holographic multiverse: our universe is mirrored on a distant brane surrounding and bounding our space, creating a parallel equivalent of our universe.

Simulated multiverse: digital creations mimicking our universe in all respects could be programmed on multiple computers.

Ultimate multiverse: every possible universe (as described by a consistent mathematical structure) is real: therefore, the multiverse encompasses all of the above.[9]

Some of these versions, fascinating as they may be, do not enter very directly into the historical multiverse story. The holographic multiverse, for instance, may be a fruitful concept for understanding fundamental issues in physics, but it doesn't really fit well into the narrative about a plurality of worlds. It's also fun to imagine that our universe is a computer simulation—perhaps a bright high school student's science fair project.[10] If it is, no doubt a multiverse of similar such simulations would exist as well. But they wouldn't illuminate the story of the multiverse for inhabitants of this universe (even if it is a simulation). And I don't believe anyone writing a book on the multiverse (or anything else) would believe they were simulated into doing it.

In any case, if there is a multiverse, the crucial point is not what kind it is. The key issue now is not the method by which other universes got created. Rather it's whether they exist at all—and, most important, whether they offer a set of different conditions or physical laws, a panoply of properties that make them useful for explaining things in the universe we live in.

As Frank Wilczek notes, modern scientists have traditionally considered the universe to be all that exists and have "tacitly assumed that the same fundamental laws apply at all places and at all times." He refers to that assumption as "universality." In that context "multiversality" would be taken to mean that different laws apply in different places (or times). That would redefine multiverse as "a physical model embodying multiversality." With those definitions, Wilczek says, the scientific question then becomes whether aspects of observable reality can be explained only by multiversality.[11]

A multiverse of identical universes would not be suitable for that purpose. And that's one of the features of the modern multiverse that distinguishes it from some of its medieval predecessors. Aristotle, while denying the possibility of other universes, made the point that if any such universes did exist, they would be identical to our own, containing the same elements, observing the same rules. Medieval debates about the plurality of worlds (conducted, typically, on Aristotelian terms) generally concurred. (The atomists of ancient Greece, though, did imagine other worlds with a variety of properties.)

Frank Wilczek

Even as the debate continued into the modern era, for the most part the plurality of worlds meant a plurality of similar worlds. Stars in other planetary systems would be more or less like the sun; island universes would all be big systems of stars like the Milky Way.

Still, a multiverse whose members differed substantially from one another had been hinted at from time to time. Nicholas of Cusa, for instance, who believed in an infinity of worlds, speculated that life would differ from world to world because of differing environmental conditions. And remember Montaigne in the sixteenth century: "Perchance," he wrote of other universes, "they have a different appearance and different laws." Later Isaac Newton, you will recall from chapter 8, saw "nothing of contradiction" in the idea of varying laws of nature in different parts of the universe. In the eighteenth century, Immanuel Kant noted that universes with extra dimensions of space would differ from ours. In 1848, when Edgar Allan Poe wrote of the universe as "clusters of clusters" of stars, he imagined the possibility of other such universes, too distant to see, with different physical laws. "*If* such clusters of clusters exist . . . it is abundantly clear that, having had no part in our origin, they have no portion in our laws," Poe wrote in *Eureka*. "Their material—their spirit is not ours—is not that which obtains in any part of our Universe."[12]

Today most multiverse investigations rely on such diversity. Some physicists would even say that the defining feature of the "cosmological multiverse" is variability from universe to universe in the apparent "local" laws and features of physics, even though these local peculiarities would all emerge from a set of universal fundamental laws. As far as we can see, the local laws operating in laboratories on Earth apply everywhere else as well, but there's no guarantee that such constancy extends infinitely. At some great distance, beyond our cosmic horizon, there may be regions of space where our lab-tested laws of physics no longer govern. "The local laws of physics could appear different in different regions—the number and masses of particles, forms and strengths of interactions between them, or even the number of macroscopic dimensions of space," writes Sean Carroll.[13]

Such diversity in local laws of physics is the hallmark of a Level II multiverse, the multiverse that many experts suspect may have been generated by chaotic inflation. In the chaotic inflation picture, as Steven Weinberg points out, energetic burblings in various portions of the early universe led to multiple big bangs, with the various Big Bang–bubble offspring differing in crucial properties. "One of the things that was burbling in the very early universe was the value of various fields, whose average value determines what we think of as fundamental constants like masses and so on," Weinberg says. "So these different regions which managed to expand into a regular big bang came with different values of what we usually think of as fundamental constants."[14] And such regions of space differing in local physics—bubbles in different "vacuum states"—fit nicely with the "landscape" of varying properties deduced from superstring theory.

NOT JUST COSMOLOGY

Differing properties from universe to universe—Wilczek's multiversality—is the essential feature needed to put the multiverse to work for scientific purposes. Having many universes with nonidentical properties is at the heart, for instance, of the use of anthropic reasoning to explain the universe we live in. Weinberg likes to reiterate his example of the Earth's surface temperature. If Earth were the only planet in the universe, or if all planets possessed precisely the same properties, then its curiously comfortable temperature (for life) would be a mystery. You might be forced to conclude that some benevolent deity had arranged things that way. But in a universe with many planets, with differing conditions, the mystery dissolves.

"Once we know that there's a very large number of planets, which have a fairly broad range of masses and distances from their parent star and chemical compositions and so on . . . then you can calculate what kind of planet you expect to find yourself on, considering that it has to be a planet on which life could evolve," Weinberg says. "And it's not that different from what we actually see." He finds it "satisfying" that this view eliminates the need for that benevolent deity. But that wasn't the motivation, he adds. "We did not invent the multiverse with atheistical motives. But it's true that once, for one reason or another, you start thinking about a multiverse, it does remove the necessity for benevolence built into the laws of nature or into the structure of the world."[15]

Weinberg and his collaborators used this sort of reasoning to calculate why our cosmos contains the Goldilocks amount of dark energy, just right for a universe that offers hospitable conditions for people (or life in general). But the multiverse provides explanatory power for more than just the mysteries of cosmology. Fundamental aspects of particle physics are also hard to explain without help from the multiverse. Even the existence of stable atoms requires what appears to be extraordinary luck without the

Steven Weinberg

explanatory power of the multiverse's anthropic implications, as physicist John Donoghue of the University of Massachusetts Amherst emphasizes.

Donoghue studies the Standard Model of particle physics. As you might recall from chapter 13 (just a few pages ago), it's the set of equations that describes the properties and behaviors of nature's basic particles and forces (excepting gravity). Among those particles are the quarks (which make up the protons and neutrons of the atomic nucleus) and leptons (such as electrons and neutrinos). Forces include electromagnetism, the strong nuclear force (which holds quarks together in protons and neutrons), and the weak nuclear force (involved in radioactivity and other subatomic processes).

The Standard Model's basic structure is determined by mathematical symmetry principles. But within that structure many quantities are not fixed by the theory, Donoghue points out. Nature is free to choose values for things like the masses of the particles and the strengths of the forces. "There are these numbers that are not predicted by the theory, that we've gone out and we've measured them—those are the parameters of the

John Donoghue

theory," Donoghue explained to me when I interviewed him in March 2018.[16]

The Standard Model has been spectacularly accurate in predicting real-world experimental results. You can see, Donoghue says, "how it produces the world." But if you changed some of its parameters, even very slightly, you get a world that is nothing like what we see.

Suppose you make the mass of the electron three times bigger. "Then all the atoms disappear, so the world is a very, very different place," Donoghue says. "The electrons get captured onto protons and the protons turn into neutrons . . . and so you end up with a very strange universe that's very different from ours."

Other aspects of the Standard Model also seem constrained by the observational fact of life's existence. One example is the mass of the famous Higgs boson. That particle is a product of the Higgs field, which permeates all of space and is responsible for conferring mass to other subatomic particles. In 1998 Donoghue and colleagues showed that the strength of the Higgs field must lie within a narrow range to produce physics compatible with life.[17] "If you change it by a bit then atoms don't form and nuclei don't form,"

Donoghue says. "You would not have any chance of having life in such a universe."

Other slight changes in the Standard Model would also destroy the possibility of structured matter. Protons, for instance, are slightly less massive than neutrons. But if you increased the mass of one of a proton's constituent quarks by just a bit, protons would become heavier than neutrons and would decay into other particles (like neutrons do). Again, atoms as we know them (and life) would be impossible.

About half a dozen Standard Model parameters need to be just so "in order to satisfy the need for atoms, the need for stars, planets, et cetera," Donoghue says. "We would like eventually that those are predicted by some other theory. But that's the question—whether they are predicted or whether they are in some sense random choices in a multiverse."

For the multiverse explanation to work, Donoghue explains, the universe needs to encompass multiple domains with different properties, or different "ground states" (the stable state a system exists in with no added energy). For Donoghue's purposes, a multiverse corresponds to the existence of many ground states—"lots and lots of them"—that are realized in different parts of the universe. A useful multiverse scenario requires an underlying theory that can accommodate not only multiple ground states, but also a physical way of producing regions that exist in those different ground states.

It's the possibility of multiple, diverse ground states that makes multiverse reasoning so powerful in explaining why the properties we see in our portion of the universe are so life-friendly. It doesn't matter how that diversity originated, as Weinberg emphasizes.

"Once you assume that there is a huge variety of universes, all of which in some sense can be said to exist with different values for fundamental constants, the consequences really don't depend very much on the details of where that came from," he told me during an interview in February 2018. As long as the fundamental theory provides for stable circumstances with different local physics, it's irrelevant whether the method for generating such a multiverse comes from inflation or Everett's branching quantum wave function.

"It doesn't pay me to try to think very hard about these alternatives," Weinberg says. "You can still use the multiverse idea to calculate what you expect, including the experimental bias that we're here to ask the question."[18]

UNITY IN DIVERSITY

Weinberg also notes that the different possible visions of the multiverse may not be independent. String theory and inflationary multiverses could be related, for instance. "It may be that the whole physics of this is described by string theory and the different expanding big bangs correspond to different solutions of string theory," he says. Or, as others have suggested, the string multiverse might somehow correspond to the many worlds of Everett's quantum mechanics.

But whether the various possible multiverses are related on a technical level or not, it's certainly true that they are not necessarily mutually exclusive. Perhaps all (or at least many) of the multiple possible versions of multiverse exist, reflecting individual manifestations of a deeper, more general conception of the universe.

Jim Hartle, who with Murray Gell-Mann developed the view of quantum mechanics called decoherent histories (as discussed in chapter 11), believes a more generalized notion of multiverse can resolve many troubling issues. By *multiverse,* Hartle means "an ensemble of alternative possible situations only one of which is observed by us." Many such types of alternative situations are possible. In fact, such alternatives are required by the basic ideas underlying quantum mechanics and its application to cosmology. If you want inflation to produce many bubbles (or "pockets") of space with differing fundamental constants, all you need is a quantum mechanical theory describing a "quantum state" of the universe and a dynamical rule specifying how that state changes over time. A quantum state can evolve into many possible sequences of events, or histories.

As Hartle and Gell-Mann describe the quantum nature of reality, we live in one set of such histories that are sufficiently similar that they appear to be just one reality. We don't notice the subtly different versions of our history, because we view the world with a certain amount of "coarse graining." Just as a street map of New York City does not record every sidewalk crack, our narrative describing reality ignores all sorts of minor details that don't affect the big picture. We take the temperature of the air without trying to follow the paths of every molecule, for instance. For all chains of events that everybody agrees on, nobody observes the precise behavior of every molecule at all times.

In this sense even our perceived reality is a "multiverse"—it incorporates a whole set of histories that differ slightly from one another, just at a level too minute to notice. But our set of consistent histories isn't the only such set. Reality can accommodate other sets of consistent histories totally separate from ours. Those other sets, each a multiverse unto itself, compose a grander multiverse, on a higher level.

Hartle contends that most versions of the multiverse can be understood in this framework. Quantum theory does not predict merely one multiverse consisting of alternative histories, but rather it predicts multiple multiverses. "A quantum system like our universe is not described by only one quantum multiverse," Hartle writes. "It is described by many different ones at different levels of coarse graining within the same theory."[19]

The inflationary multiverse, consisting of many spacetime bubbles—or "pocket universes"—is one of the many types of multiverse that quantum processes can generate. A vast spacetime containing many pocket universes, each with different values for fundamental constants, is produced automatically by quantum processes describing an eternally inflating false vacuum. "Quantum multiverses are not some posited speculative idea to be grafted onto the basic theory. They are the output of that theory," Hartle declares.[20]

So in Hartle's picture, reality is a multiverse both in the sense that it includes many ensembles of coarse-grained consistent histories (Everett's many worlds) and in the sense that multiple bubbles inflate out of the quantum vacuum. In each set, only one history is observed by its inhabitants, with a probability that can be calculated from the quantum math.

Hartle argues that viewing multiverses as arising from sets of coarse-grained quantum histories refutes many of the objections expressed by multiverse critics. In particular, this picture should allay concern about using the anthropic principle to explain features of our universe. In fact, there is no need to propose any special anthropic principle at all. The anthropic argument—that conditions must be amenable to our existence—emerges automatically from the quantum calculations of probabilities. "Anthropic reasoning does not rely on some anthropic principle and is not a choice to be made or not made," Hartle writes. "It is an automatic consequence of calculating probabilities for observations."[21]

Wilczek applies similar reasoning when using Everett's many worlds to explain the notorious indeterminism of quantum mechanics. The results of quantum measurements cannot be predicted precisely, only statistically. Yet the probabilities for different outcomes are forecast by a precisely deterministic equation (the wave function). How indeterminism can emerge from determinism is at the heart of many arguments over how to interpret quantum mechanics. But the quantum multiverse is exactly what solves this mystery, Wilczek explains.

"The probabilities assigned to different outcomes, in this interpretation, describe the probabilities for finding oneself on the branch of the wave function, or 'world,' where the given outcome has occurred," he writes. "The wave function as a whole evolves deterministically, but it describes a multiverse, only one part of which remains accessible." Universality cannot accommodate these multiple quantum outcomes. But a multiverse can. So when Wilczek asks if any aspects of observable reality can be explained by multiversality, but not otherwise, he can answer yes: "one is the apparent indeterminism of quantum mechanics, despite its deterministic equations."[22]

Hartle's formulation of the multiverse idea addresses another common complaint: that the multiverse is not observable or testable and is therefore not science. He rejects that allegation. Just because we can't run a sufficiently large experiment and wait billions of years to see multiple big bangs doesn't mean that multiple inflationary universes don't exist. "The laws of the universe do not have to be such as to make it easy for some negligible bits of protoplasm to test them directly on the scales that happen to be accessible to them at the moment," Hartle writes.[23]

Similarly, we can't see the other "worlds" prescribed by the Everett interpretation of quantum mechanics. But we can infer that they exist just as we infer that evolution occurred to produce the various life-forms that we do observe. Evolution could have proceeded in many ways; different possible histories represent different branching trees among the possibilities. In the same way, we observe one of many different branches of possible quantum histories for the universe (or multiverse) we live in. "We do not observe directly ('see') any of the histories of this multiverse," Hartle points out. "They are in our past. Yet we believe that biological evolution happened because it explains regularities today."[24] Similarly the multiverse (or multiverses) explains many of the regularities we see today in nature as a whole.

MULTIVERSE DENIERS

While a multiverse (of various sorts) is capable of explaining many phenomena, it remains true that many scientists today don't like the idea of the multiverse and consider it to be wrong, not science, or both. "There are very good scientists—I don't mean that these are people outside of science, these are very good scientists—who think the multiverse idea is unscientific," Weinberg says. But he does not agree. The situation with the multiverse is much like that with string theory, also alleged by some to be not science. String theory posits unobservable things (the strings), but is nevertheless a scientific theory—just one that is not yet complete and established to be correct. "String theory hasn't had enough successes so that we know it is right and we don't even know in detail what it is," Weinberg says. "But there's nothing unscientific about string theory. We'll never observe the individual strings—so what? If it makes a lot of predictions that we can test, then you're home free. And likewise with the multiverse."[25]

Nevertheless, resistance to the multiverse remains considerable. To a large extent, that resistance reflects hostility to anthropic reasoning, as discussed in chapter 12. But there are several additional lines of attack on the multiverse, which for the most part fall into three broad categories:

1. It is not science (or it is wrong) because it violates Ockham's razor, which demands explanatory simplicity.
2. It is not science because the multiple universes are not observable.
3. It is not science because it does not make testable (or falsifiable) quantitative predictions.

Ockham's razor might be the most commonly expressed objection to the multiverse, but it is certainly the weakest of the anti-multiverse arguments. For one thing, Ockham's advice itself is commonly misstated (as mentioned in chapter 5). He never said "the simplest explanation is correct" or anything like that. Another common way of expressing it—entities must not be multiplied beyond necessity—is not Ockham's phrasing, either. (Ockham's actual expression: "It is futile to do with more things what can be done with fewer.") Whatever the formulation of the principle, the key idea is that you shouldn't postulate more entities than you need to explain what you want to explain. As Newton said in his first rule of reasoning in natural philosophy, "Admit no more causes of natural things than such as are both true and sufficient."[26] If you can explain all that you see with one principle, that's great, Ockham would approve. But note Newton's "true" and "sufficient." A simple explanation is not valid if it's not true; a complicated answer may be required if a simple answer is not sufficient to explain what needs explaining. If you need more principles, or entities, to explain what you see, then you need to consider them. If you can explain the observable universe without a multiverse, then motivation for investigating the multiverse diminishes. But if multiple universes are necessary to explain what you see, it is not a violation of Ockham's razor to include them in the scientific conception of reality. (And don't forget that Ockham himself favored the existence of a plurality of worlds.)

Many of the people who have cited Ockham's razor and expressed other common complaints have not thought them through very carefully. But a few have made a serious attempt to make the antimultiverse case in some depth. The most thorough and thoughtful critique comes from cosmologist George F. R. Ellis, who has articulated the above complaints in considerable detail.

Ellis rejects the argument that the multiverse is implied by the "known" physics underlying inflationary cosmology. Known physics is involved, he agrees, but it is applied to unknown circumstances. "The key physics . . . is extrapolated from known and tested physics to new contexts; the extrapolation is unverified and indeed is unverifiable; it may or may not be true," Ellis asserts.

He repeats the objection that the multiverse idea is not testable because the other universes are beyond the cosmic horizon and cannot be observed.[27] He recognizes, though, that science does permit the possibility of the "existence of unseen entities." But such entities can be considered real, he says, only if they meet certain criteria: "They form an essential link in a chain of argument with well supported foundations, they have demonstrable predictable effects on the physical world of matter which we see around us, that are in agreement with experimental phenomena, and there is no other plausible explanation for the same phenomena." He argues that multiple universes don't meet these tests, because the foundations aren't well supported (the theory of inflation has not been proven), outcomes of the multiverse hypothesis are not predictable, and you could explain everything that the multiverse explains just by "sheer chance."[28]

Ellis does acknowledge claims that postulating a multiverse has in fact produced a prediction: the quantity of the dark energy permeating the universe. "But these are statistical predictions," Ellis points out. "They have no meaning if there is only one universe. They . . . cannot be taken as proofs, for they take for granted what is to be proven."[29]

All in all, Ellis dismisses the multiverse as hypothetical, unsubstantiated, and a departure from rigorous scientific standards. "Once the gold standard of experimental support has been dropped and basic scientific principles are disregarded, why should such theories be regarded as good science?" he writes.[30]

Taking his arguments at face value, Ellis makes a strong case that the multiverse is not scientifically established as a true and correct description of reality. The problem is that nobody says it is. Weinberg's dark energy prediction has never been suggested as a "proof" of the multiverse, not even by Weinberg. "I wouldn't say that the successful anticipation of the fairly large dark energy is enough of a success to confirm the multiverse idea," Weinberg told me. "It's a slight indication that we should take it seriously."[31]

Ellis himself states that the extrapolation of known physics to the multiverse context "may or may not be true." And that's just the point. It may be true. That's what scientists studying the multiverse idea are trying to find out—whether it is true or not. After all, while the Standard Model is well-established physics, correct in every way it can be tested, it doesn't explain everything. Physicists must explore other theoretical approaches to explain what the Standard Model can't, and the multiverse is one possibility. It's illogical to claim that the multiverse is not a valid approach to answering today's unanswered questions because theories predicting it are not well-established. "That's true of all theories beyond the Standard Model," Donoghue points out. "None of them are established yet. So we can't really say with any confidence that there is a multiverse. It's a physical possibility. It may be wrong. But it still may be right."[32]

While Ellis admits that some unobservable things do exist, he claims that the multiverse does not meet all his criteria, because it's possible that what those other universes explain could also be explained by "sheer chance." But that argument essentially invalidates his criterion, or else it eliminates a lot of other science, because you can explain virtually anything by invoking sheer chance. If you allow that option, you can deny the reality of any unobserved entity whatsoever.

Besides, as Wilczek has pointed out, criteria such as observability (or the simplicity of Ockham's razor) are not in themselves laws of doing science, but merely guidelines useful for evaluating scientific proposals. Saying that "nothing that is not observed exists," Wilczek says, is a statement with "no empirical content."

"It might be that the laws we use successfully to describe the observable universe are most naturally formulated in a larger framework, that includes unobservable parts," he writes. Insisting on observability as a requirement for existence is a sort of "moral exhortation, or methodological principle, not unrelated to Occam's razor . . . whose application, though usually appropriate, might be outweighed by other scientific considerations."[33]

As for the lack of testable predictions made by multiverse theories, Carroll has often pointed out that the multiverse is not itself a "theory" that is supposed to make predictions—it is itself a prediction made by other theories. "Some people say that the job of science is to come up with a model that makes predictions that fit the data, but their mistake is to think that the multiverse is a model that makes predictions. It's not a theory, it's a prediction of a theory," Carroll told me in 2006. "I find it remarkable on both sides that people object to these theories and say that it's religious or philosophy, apparently forgetting that every theory ever invented makes untestable predictions."[34]

More recently, Carroll has reiterated the important point that an unobservable multiverse can explain observable things. "The ultimate purpose of a theory is to account for what we observe," he writes. Inability to observe the multiverse does not mean that it doesn't exist, and it may be necessary to acknowledge its existence in order to explain observable features of the universe. "There really might be a multiverse out there," Carroll asserts, "whether we like it or not."[35]

NOT JUST METAPHYSICS

Ellis dismisses Weinberg's use of the multiverse in predicting the dark energy's value because it is a statistical prediction that therefore presumes the existence of a multiverse to begin with. But that razor cuts both ways. Of course Weinberg's prediction was made by assuming the existence of the multiverse. That's how you find out what the multiverse implies about observable reality. If you disallow this prediction, then you are presuming the opposite, that there is only one universe. By assuming a single-universe reality, you are specifying the answer to the question in advance, not investigating it.

Insistence on a single universe is no doubt related to the longstanding motivation in physics to find a theory that predicts all the fundamental features of nature from a set of rigorous mathematical formulas (such as string theory). "Probably literally everyone's hope is that we would someday find a theory and all of a sudden everything would become clear, there would be one unique possibility . . . there would be no choice but this was the theory—everyone would love that," says Donoghue.[36]

But as Wilczek says, it hasn't worked out that way.

"There are plenty of things that are susceptible to powerful reasoning and will tell us about really fundamental aspects of the interactions, and plenty of things to keep mathematical physicists employed and philosophically minded physicists bemused," he told me in 2006. "But I do think the line's going to move. I think we'll discover that the idea that string theory would provide you this algorithm, this machine, that all you do is turn the crank and out pops the Standard Model and all the parameters and so forth, to me that was never very credible. And now it's less credible than ever, and we're finally facing up to it."[37]

Weinberg, for one, is not at all happy with the prospect that no theory can explain basic features of the physical world. "I don't like it either," he said. But that may be the way it is. "There are lots of conditions we impose on our theories, that they have a certain logical rigidity, that they're not just full of arbitrary features," he says. "But we should not impose the condition that the theory makes us happy."[38]

Donoghue takes a similar view. "Just because you feel bad about the multiverse, and just because some aspects of it are beyond our reach for testing, doesn't mean that it's wrong," he says. "To say that the multiverse is not science is itself not science. You're not allowing . . . a possible physical theory that you're throwing out on nonscientific grounds."[39]

As things stand, the modern multiverse is certainly not regarded as settled science. Inflationary cosmology, while very successful in many respects, has not been verified beyond all plausible doubt. Superstring theory remains unverified, and the string landscape—cited as strong support for a multiverse—might or might not survive further theoretical developments. So the multiverse today is in much the same situation as "island universes" were in the first two decades of the twentieth century—a possibility embraced by some but rejected by others. But of course, that was also a time when cosmologists didn't even know about the Big Bang and the expanding universe, well-established science today. "Until 50 years ago we didn't have firm evidence for the Big Bang," Martin Rees, the British astronomer royal, said in a public lecture in January 2017. "Now we know lots about its properties and how it's led to the cosmic panorama of which we are a part." So 50 years from now, he suggested, there could be a theory that lends similar credibility to the idea of a multiverse.[40]

Rees recalled a conference where attendees were asked to rate their confidence in the existence of a multiverse, to be expressed in the form of a wager. What would you bet that a multiverse exists: your goldfish, your dog, or your life? "I was dog," Rees said. Andrei Linde, on the other hand, was willing to bet his life. Later Weinberg heard of these wagers and rejoindered that he would be willing to bet both Martin Rees's dog and Andrei Linde's life.

"Andrei Linde, my dog, and I will all be dead long before this is settled," Rees said. "But it's not just metaphysics. It's exciting science, and it may be true."

Epilogue

Now bend thy mind to truths profounder still:
For stranger doctrines must assault thine ear,
And a new scene of wonders yet unfold.
Whate'er is new, though obvious and defined,
Gains not an easy credence; but when once
Flies the fresh novelty, th' unsteady soul
Yields its full faith to facts mysterious most.

—LUCRETIUS

FROM PRIMITIVE MYTHOLOGY to modern cosmology, poets, philosophers, and scientists have all attempted to grasp the totality of existence.

Many of them intuited, or reasoned, or speculated that all existence—the universe—is vastly greater than what's accessible to human awareness. Leucippus, Democritus, and Epicurus envisioned countless versions of the cosmos created by atoms far outside the confines of the celestial spheres. In the late Middle Ages Nicholas of Cusa spoke of worlds "without number" in an essentially infinite space, as did Giordano Bruno in the sixteenth century. A few decades after Bruno, René Descartes proclaimed the "number of the heavens" to be indefinite and advised that imposing a limit on the extent of the world failed to acknowledge the limitations of human reason. In the mid-eighteenth century, Immanuel Kant imagined a "progressive hierarchy of worlds"; a century later

Edgar Allan Poe fancied that "there does exist a limitless succession of Universes." Ormsby MacKnight Mitchel spoke of thousands of "mighty Universes scattered through space," more than there are stars in the Milky Way. As the twentieth century came to an end, the astronomer royal of Great Britain, Martin Rees, declared that our universe may be as just "one atom in an infinite ensemble: a cosmic archipelago." A multiverse.

Historically such declarations of unseen cosmic vastness have invariably turned out to be essentially correct. But analogy never confers certainty. Today the multiverse remains controversial. Nevertheless the current multiverse debate echoes those of the past in numerous ways. Similar issues and arguments constantly arise as scientists write, in real time, a new chapter in the long story of the human quest to understand the universe—and to discover whether there is more than one.

Thus the multiverse is not a uniquely modern idea, born of twentieth-century advances in physics and cosmology, but is rather an ancient idea with a rich and instructive history. That history, I think, implies some important lessons not only about science's effort to understand the cosmos, but also about the nature of science itself. Most trivially, perhaps, is the lesson that however far-fetched the idea of multiple universes might be, such ideas have often eventually been validated, so similar proposals today ought not to be casually dismissed. But more generally, I think the story of the multiverse makes three deeply significant points worth a bit of elaboration:

One, *universe* is not a word with a clear, unchanging meaning. It may mean "everything," but how "everything" is conceived and described depends on the state of knowledge at any given time. As that knowledge advances, the concept of "universe" must sometimes be redefined—and that has indeed happened several times in scientific history.

Two, modern cosmology is still in some respects medieval. Nobody denies that vast progress has been made in the centuries since Robert Grosseteste, Albert the Great, and William of Ockham. But big, deep questions remain unanswered. And many of them are precisely the same questions that medieval scholars pondered and debated.

Three, how science is properly done, and even what it means to be scientific, both depend on the nature of the universe we inhabit. Living in a multiverse would imply different rules for doing science, and different criteria for what counts as science, than those appropriate for a singular universe.

CHANGING DEFINITIONS OF UNIVERSE

In antiquity and the Middle Ages, *universe* and *world* were synonyms. All that existed—the world—was a set of nested spheres, Aristotle said, with Earth in the middle. Some medieval philosophers realized that if other worlds did exist, the concept of universe would need to be rethought. Copernicus then redefined the world to be a solar system, with the sun in the middle. With the Earth now one of several planets, and the sun at the center of the world, scholars began to suspect that the fixed stars were also "suns," the centers of countless other worlds. The universe became something much bigger than it had been before—it was a universe of stars. Then Thomas Wright and Immanuel Kant proposed that this universe of stars—the Milky Way galaxy—was not alone. Many other such clusters of stars, disguised as cloudy nebulae, populated the cosmos. In the twentieth century, Edwin Hubble confirmed that these nebulae were in fact galaxies, then redefined the universe as the vast expanse of expanding space that contained them.

Today *universe* usually refers to this expanding spacetime bubble, the output of the Big Bang "explosion" 13.8 billion years ago. But there may have been more than one big bang, and so there may now be many such bubbles. *Multiverse* is the provisional term to describe the possibly infinite number of bubbles that may in fact be the true, whole universe.

A multiverse could be realized in other ways as well. Perhaps the known universe (whether one bubble or many) has counterparts in unknown dimensions of space, kind of like Twilight Zone twins (technically called brane worlds). Or perhaps multiple universes exist in the same space, as branches of quantum possibilities that all become real in one or another chain of "consistent histories." The point is that the universe we know may not be alone, and so the word *universe* itself may once again, someday, need redefining.

MODERN SCIENCE IS MEDIEVAL

From 1277, when the Bishop of Paris permitted—or actually required—philosophers to contradict Aristotle, scholars have sought answers to questions that Aristotle had considered closed. Among the most prominent of those questions was the possible existence of a plurality of worlds. Leading experts debated whether the universe as then conceived was solitary or accompanied by cosmic siblings. Today the debate about the multiverse is similarly framed. Some cosmologists advocate for a multiplicity of universes, others reject the idea—just as medieval philosophers disagreed about a possible plurality of worlds.

Current cosmology mirrors medieval natural philosophy in several additional ways. Both medieval and modern cosmologists developed a view of the universe rooted in a mix of theory and observation, for instance. Medieval philosophers conceived a cosmos based on Aristotle's logic of natural motion and place, plus observations of an apparently rotating sphere of stars. Today's view of the cosmos is based on Einstein's mathematical logic describing space and time, plus observations of the apparent expansion of the universe.

Another question encountered by both medieval natural philosophers and modern scientists is whether the universe is eternal or had a beginning. Aristotle had argued strongly for eternal. But that conflicted with the Christian creation story. Many medieval philosophers (such as Robert Grosseteste) sought rationales for reconciling Christian creation with Aristotle's cosmology. In the twentieth century, modern cosmology's creation story—cosmic inflation—raised a similar question. Our universe may have inflated into existence from a preexisting space—which may itself have been the offspring of an earlier "universe." Most inflation experts believe that there must have been a beginning to the inflationary process.[1] But the issue is not definitively settled, and Leonard Susskind has proposed that the cosmic past could extend back to "minus infinity," a view closer to Aristotle's.[2] The medieval question about a past eternity remains unanswered.

Medieval philosophers also pondered the ancient Greek view that "celestial" matter in the heavens differed in nature from matter making up the Earth. They debated just what the properties of that celestial matter might be—whether it was crystalline and rigid or fluid, for example. Today scientists know that the bulk of cosmic matter does, in fact, differ from anything found on Earth. They label it "dark matter," but nobody can say just what it actually is.

Other similar questions appear in both the medieval and modern context. Can a void exist beyond the universe we inhabit? Aristotle said no, but some medieval scholars contended otherwise. Today the known universe is considered to be largely a vacuum. Then again, some physicists picture the universe's three space dimensions as fitting within an empty "bulk" space of higher dimensions—a different kind of void, that may or may not exist. Nobody really knows.

More generally, medieval experts debated whether science should restrict itself to direct experience, or might also use reason to deduce the existence of unseen factors, abstracted from experience. Averroës, the Muslim commentator on Aristotle, "identified the real world with the directly observable and the concrete." Averroës argued that abstract concepts were imposed on nature by modes of human thought. But others, such as Avempace (c. 1085–1138), suggested that a deeper reality could be revealed from idealizations drawn by reason.[3] A similar argument is alive in science today, as some physicists argue that the multiverse, being not observable directly, is not science at all, while others permit the inference of unseen things by reasoning about things that can be seen.

Today's arguments, of course, take place in the context of a much deeper understanding of the cosmos than natural philosophers possessed in the days of Averroës and Avempace. Everybody acknowledges that science has made enormous progress since the Middle Ages. Modern science has mastered the microworld of atoms and molecules, discerned the secrets of stars and planets, unlocked many mysteries about the Earth's atmosphere and its innards, perceived the mechanisms of life and the origin of its multiplicity of species—not to mention the architecture of the human body and brain. And modern cosmologists have certainly grasped the vastness of the cosmos in detail unthinkable by their medieval predecessors.

And yet for all their successes, scientists today struggle with issues very similar to those that occupied the best medieval minds. Profound questions remain unanswered, deep mysteries unsolved. So science's current efforts to discover truths about the workings of the natural world remain in a sense medieval—that is, a prelude to an even deeper understanding that may not come for another millennium.

History suggests that today's grandiose claims of scientific success should be tempered by an appreciation of how current science is likely to be viewed in the future. A few centuries from now, today's grand scientific edifice might very well be viewed as something like a medieval cathedral—magnificent, but nonetheless a product of a primitive intellectual age. Some of today's big questions will surely be answered by then, but some not, and some will be seen as insignificant, or at least improperly posed. In any case, progress toward answering those questions will no doubt depend on how the practice of science itself evolves.

HOW SCIENCE IS PROPERLY DONE

At the core of the multiverse story is the lesson that the proper way of doing science depends on the nature of the reality within which science is practiced. Rules for doing science in a vast expanse of spacetime alike in all respects, in all times and places, would differ from rules that are appropriate if multiple such spaces exist and are not all alike (whether observable or not).

Take Kepler's approach to explaining the solar system, mentioned in chapter 12. He attempted to determine why there were six planets, spaced about the sun as they were. He believed that he could deduce the architecture of the solar system from fundamental mathematical principles. He thought he succeeded! By nesting the five regular ("Platonic") polyhedral solids within one surrounding sphere, he described a geometry with six spaces for planets. By assigning the proper order for the solids, the spacing came out more or less right, too. But he was wrong.

"The model was impressive because Kepler used mathematics to explain observed phenomena," comment Mario Livio and Martin Rees. "It was completely wrong because Kepler didn't understand at the time that neither the number of the planets nor their orbits were fundamental phenomena that required first-principles explanations."[4] Today we know that the number of planets and locations of their orbits are not fixed by any mathematical theory—they are accidents, determined by many environmental details as a planetary system forms. The universe contains numerous planetary systems, none completely alike.

But suppose Kepler lived in a universe with only one sun and only six planets. In that case, an "environmental accident" would not have explained the locations of the planets with respect to the sun—there would be only one, unvarying environment. His approach to finding a scientific explanation might in that case have been sensible and proper, and maybe even right. Similarly, scientific methods that work well in a one-shot universe might not be so appropriate if we actually live in a multiverse.

This consideration applies to the general view that science is supposed to provide mathematical predictions based on quantitative laws. Like Kepler, many modern scientists are obsessed with finding mathematical rules that prescribe the properties of the universe. A deeply ingrained tenet of modern physics is the belief that the fundamental features of nature should be explained as the consequence of basic principles, expressed in mathematical form, providing precise predictions. To be scientific, these predictions must be testable by experiment. In philosophical jargon, a scientific theory must be "falsifiable"—its predictions must be testable in such a way that it's conceivable they could turn out to be wrong. Many multiverse opponents argue that the multiverse does not meet this test.[5]

Among the most outspoken in the regard is Paul Steinhardt, one of the early investigators of cosmic inflation. He now rejects it. Steinhardt's lament seems to be that eternal inflation would generate all possible universes with all possible properties. It's therefore no mystery that it explains our universe, because it could explain any universe.

"If you ask what inflation predicts, the safest thing to say is just that anything that can happen will happen and will happen an infinite number of times," Steinhardt says. "So what we have is really, I would say, an unmitigated disaster as a theory, a theory which is maximally unpredictive. Literally any physically possible cosmic property will occur in this multiverse . . . so there's no way to falsify this theory, or verify it for that matter, because anything could fit."[6]

Steinhardt's antimultiverse argument is yet another echo of reasoning from the past that turned out to be unmitigatedly wrong. Historian Helge Kragh has noted that the founders of the steady-state theory of the universe made a similar case against the expanding Big Bang universe based on Einstein's general theory of relativity. "They were struck by the lack of uniqueness in the relativistic expansion models," Kragh wrote, "and felt repelled by the fact that these models could accommodate almost any observation, and hence had little real predictive power."[7] It turned out, of course, that Big Bang cosmology was the fruitful approach, and the steady-state universe is now largely forgotten. Steady-state advocates were arguing against a work in progress. Eventually the Big Bang theory did make predictions that were confirmed.

Similarly, there may in fact be some observational tests for the multiverse. A neighboring bubble might bump into ours, for instance, and produce patterns on the cosmic microwave background radiation, the cold glow of microwaves left over from the Big Bang. Some peculiar features in the microwave background have actually been reported, and some scientists have suggested they are possible signs of other universes, but that interpretation has not yet received widespread endorsement.

It has been well-established, however, that temperature fluctuations in the microwave background reflect quantum processes during the epoch of inflation. Success in calculating the observed properties of the cosmic microwaves suggests that other implications of those quantum fluctuations should not be disregarded. In almost all versions of inflation that cosmologists have studied, those quantum fluctuations lead to eternal inflation and an infinite multiverse, Alan Guth points out. "Given our success in calculating the fluctuations observed in the cosmic microwave background, it should make good sense to consider the other consequences of quantum fluctuations in the early universe," he asserts. "I think it is time to take the multiverse idea seriously, as a real possibility."[8]

Other tests of a multiverse are possible. John Donoghue has proposed that the distribution of subatomic particle masses would differ in recognizable ways depending on whether a multiverse exists or not, and Frank Wilczek has made a similar point.[9]

But the possibility of predictions misses the bigger issue of whether such predictions are necessary at all. In a multiverse, demanding that a theory predict all the features of reality that can be observed is a methodological mistake. As Brian Greene notes, "Asking for such an explanation is asking the wrong kind of question; it's invoking single-universe mentality in a multiverse setting."[10]

When I interviewed him in 2006, the late Joe Polchinski expressed similar sentiments. "There's this question, is it science if it's not predictive?" he said. "And it's a good question. But we don't know a priori how predictive science is. We don't know which things are absolutely predictable and which things are not, and we have to figure out the answer to this to the best of our ability." If some things turn out to be unpredictable, he said, then that's the way it is. "Arbitrary definitions of what science is shouldn't affect one's attempts to figure out the way things really are."[11]

Science, in other words, is not a preordained process governed by never-changing methods and rules. It's a quest to explain the world. And as Steven Weinberg has emphasized, part of that quest is finding out how to go about it.

"The sort of thing you should do in doing science was not clear at the beginning, and we've gradually learned it because we prize the satisfaction we get when we succeed at explaining something," Weinberg told me. "We're in the grips of a teaching machine, the world, which gives us little rewards and pellets of satisfaction when we get it right. But part of getting it right is to know what questions to ask and what are the standards that you apply to judge the answers."[12]

So when the multiverse deniers are saying it's not science, they are presupposing a definition of science that rules out multiverses to begin with. And that's not scientific.

EXHILARATED BY THE CHALLENGE

Those who say the multiverse is not science, because it's not testable, or because those multiple universes cannot be observed, are assuming the universe to be the way they think it is (or want it to be), not seeking the true nature of reality. The correct rules for doing science—for understanding the universe—depend on what the nature of reality actually is. Aristotle's methods might well have been the correct way to do science, had he really lived in a universe consisting of ethereal spheres carrying heavenly bodies in orbit around a stationary Earth.

His science was built on a combination of logical deduction, uncontrolled observations and a few foundational principles, such as the symmetry of perfect circular natural motion in the heavens and linear natural motions for the terrestrial elements. You could argue that his methods contained conceptual flaws. But then modern science's methods are not exactly flawless, either (just look at the current concerns over irreproducible results of experiments using statistical methods for hypothesis testing).[13] Newton also described a universe using methods now regarded as rigorous and modern, but his universe does not really resemble in any detailed way the current Big Bang cosmos governed by relativity and quantum physics. Newton's science would have been fine, perhaps, if he lived in a universe conforming to the laws of classical physics. But it turned out that reality hadn't read Newton's *Principia*. So five centuries from now (or four or ten) scientists might very well refer to today's cosmology as medieval, and regard Newton in much the way that we regard Aristotle.

Nevertheless, it is true that nobody really knows whether the multiverse picture of reality is correct. Inflationary cosmology and string theory (or M theory) may both predict a multiverse, but neither of those theories is yet established to be correct. They might turn out to be wrong. Quantum mechanics is supremely well-established, and it appears to predict a type of multiverse, but that interpretation is not universally shared.

So maybe there are no other universes, no other quantum worlds, or no other extra dimensions for multiple branes to occupy. But maybe there are. And maybe we'll never know. In a public talk in 2003, Weinberg expressed a pessimistic view. "I think it's one of the tragedies that face the human species that there will always be a mystery about the universe," he said. "It may not be a mystery like any of the ones that we face today. It may be that all the questions that bother us today like the origin of the Big Bang and so on will be answered. But I think there always will be some irreducible mystery, and it's just one of these tragedies we have to live with."[14]

Even so, there will always be cosmic explorers seeking the answers to the questions that remain, however difficult they might be.

"Some of us are exhilarated by the challenge of it," Leonard Susskind said when I interviewed him in 2006. "Some of us like danger. I don't like physical danger very much, but I love the danger of an idea that's very, very hard and very, very difficult— where it might take 500 years to figure out. . . . I think some people get depressed by something that looks like it's not going to be resolved for 500 years. But maybe, just maybe, somebody will come along and figure it out next year."[15]

So far, nobody has. The multiverse debate continues.

AFTER INVESTIGATING THE HISTORY of this issue, I can find no justification for ruling out multiple universes in principle. Denying the possibility of a multiverse ignores Descartes's exhortation to "beware of presuming too highly of ourselves" by supposing that there are "limits to the world" we are capable of correctly imagining.[16] As James Hartle notes, we are "negligible bits of protoplasm" in comparison with the scales of size and time characteristic of the cosmos. It therefore seems reasonably plausible that there's more to the universe than we know, possibly more than we can ever know. And so to me, it makes much more sense for a multiverse to exist than not.

Of course, as Huygens said, "I can't pretend to assert any thing as positively true (for that would be madness)," only "advance a probable guess" for others to examine.[17] Maybe there's a multiverse out there, maybe not. Nevertheless, I think it's a good bet that the immensity of the cosmos extends beyond the grasp of human comprehension. That immensity gave Kepler nightmares but instilled in Fontenelle more freedom to breathe, and to envision a universe "incomparably more magnificent than it was before." I'm with Fontenelle.

And also with Huygens, when he expressed the true mystery of science and the debates it engenders about the nature of reality. "Those sciences which are now in debate, are so much the more difficult and abstruse, that their late invention and slow progress are so far from being a wonder, that it is rather strange they were ever discovered at all."[18]

Notes

Illustration Credits

Index

Notes

INTRODUCTION

V. Agrawal et al., "Anthropic Considerations in Multiple-Domain Theories and the Scale of Electroweak Symmetry Breaking." *Phys. Rev. Lett.* 80, no. 9 (March 1998): 1822–1825.

1. Tom Siegfried, "Success in Coping with Infinity Could Strengthen Case for Multiple Universes," *Science News* 175, no. 12 (June 6, 2009), 26.

2. Astronomers also warn that the explosion analogy should not be taken too far. While the Big Bang was similar to an explosion in many ways—creating a huge fireball, for instance—it was not as if preexisting material was ignited and blasted outward.

3. Sean Carroll, "Beyond Falsifiability: Normal Science in a Multiverse," January 15, 2018, 5, arXiv:1801.05016.

Chapter 1. OUT OF CHAOS, A MULTIVERSE

David Rittenhouse, "Oration," in *The Scientific Writings of David Rittenhouse*, ed. Brooke Hindle (New York: Arno Press, 1980), quoted in Michael J. Crowe, ed., *The Extraterrestrial Life Debate: Antiquity to 1915* (Notre Dame, IN: University of Notre Dame Press, 2008), 215.

1. Tom Siegfried, "The Shape of Things to Come," *Dallas Morning News,* June 22, 1998.

2. Lawrence Krauss, interview with author, April 28, 2001, Washington, DC.

3. Saul Perlmutter, presentation at the meeting of the American Astronomical Society, Washington, DC, January 8, 1998.

4. John Archibald Wheeler, *A Journey into Gravity and Spacetime* (New York: Scientific American Library, 1990), 11–12.

5. Einstein's key general relativity equation can be written in several ways; one common form is $R_{\mu\nu} - \frac{1}{2} g_{\mu\nu} R = \frac{8\pi G}{c^4} T_{\mu\nu}$. The symbols on the left of the equal sign describe the geometry of spacetime; those on the right

correspond to the mass-energy density. With the addition of the cosmological constant term lambda, the equation becomes

$R_{\mu\nu} - \dfrac{1}{2} g_{\mu\nu} R - \Lambda g_{\mu\nu} = \dfrac{8\pi G}{c^4} T_{\mu\nu}$. For a more technical yet relatively

accessible description of the equation see David D. Reid, Daniel W. Kittell, Eric E. Arsznov, and Gregory B. Thompson, "Review of the Standard Presentation of Cosmology," in "Our Universe: A View from Modern Cosmology," October 16, 2002, https://ned.ipac.caltech.edu /level5/Sept02/Reid/Reid2.html.

6. Albert Einstein, "Cosmological Considerations on the General Theory of Relativity," *Sitzungsberichte der Preussischen Akademie der Wissenschaften* (1917), in *The Collected Papers of Albert Einstein,* English Translation Suppl., vol. 6: *The Berlin Years,* ed. Martin J. Klein, A. J. Kox, and Robert Schulman, 421–433 (Princeton: Princeton University Press, 1997), 432.

7. While Einstein's "solution" seemed sound on paper, it would not have worked in the real universe. His cosmological constant did provide a description of a static universe, but it was not a stable solution. Everything would have to be precisely balanced; a slight disturbance in the matter density would drive the universe away from its static state.

8. Helge Kragh, *Cosmology and Controversy* (Princeton: Princeton University Press, 1996), 30.

9. I first encountered this story in Hubert Reeves, *The Hour of Our Delight* (New York: W. H. Freeman, 1991), 86–87.

10. Alan Guth, *The Inflationary Universe* (Reading, MA: Addison-Wesley, 1997), 184–185.

11. Guth, *Inflationary Universe,* 191.

12. Andrei Linde, "Nonsingular Regenerating Inflationary Universe," July 1982, http://www.andrei-linde.com/articles/nonsingular-regenerating -inflationary-universe-pdf.

13. Linde, "Nonsingular Regenerating," 6.

14. Andrei Linde, "A Brief History of the Multiverse," December 3, 2015, rev. December 8, 2017, 1, 5, arXiv:1512.01203v1.

15. Robert Kirshner, news conference, meeting of the American Physical Society, Columbus, OH, April 17, 1998.

16. Robert Kirshner, presentation at The Missing Energy in the Universe conference, Fermilab, May 1, 1998.

17. Michael Turner, presentation at The Missing Energy in the Universe conference, Fermilab, May 1, 1998.

18. Josh Frieman, "Opening Remarks," at The Missing Energy in the Universe conference, Fermilab, May 1, 1998.

19. Turner, presentation at Fermilab.

Chapter 2. ROBERT GROSSETESTE'S MULTIVERSE

Doctor Strange, directed by Scott Derrickson, Walt Disney Studios Motion Pictures, 2016.

1. A. C. Crombie, *Robert Grosseteste and the Origins of Experimental Science* (1953; Oxford: Clarendon Press, 1970), 10.

2. David C. Lindberg, *The Beginnings of Western Science*, 2nd ed. (Chicago: University of Chicago Press, 2007), 393.

3. Crombie, *Robert Grosseteste*, v.

4. Francis Seymour Stevenson, *Robert Grosseteste, Bishop of Lincoln* (London: Macmillan and Co., 1899).

5. In addition to Stevenson's biography and Crombie's *Robert Grosseteste and the Origins of Experimental Science*, useful biographical information on Grosseteste can be found in Neil Lewis, "Robert Grosseteste," *Stanford Encyclopedia of Philosophy*, Summer 2013 ed., May 8, 2013, https://plato.stanford.edu/archives/sum2013/entries/grosseteste; A. C. Crombie, "Robert Grosseteste," *Dictionary of Scientific Biography*, vol. 5, ed. C. C. Gillispie (New York: Charles Scribner's Sons, 1972); Daniel Callus, "The Oxford Career of Robert Grosseteste," *Oxoniensia* 10 (1945): 42–72, http://oxoniensia.org/volumes/1945/callus.pdf; and James McEvoy, *Robert Grosseteste* (Oxford: Oxford University Press, 2000).

6. McEvoy, *Robert Grosseteste*, xiv.

7. McEvoy, *Robert Grosseteste*, xiv.

8. Lindberg, *Beginnings*, 27.

9. Michael J. Crowe, *Theories of the World: From Antiquity to the Copernican Revolution*, 2nd ed. (Mineola, NY: Dover, 2001), 9.

10. For biographical information on Eudoxus, see G. L. Huxley, "Eudoxus of Cnidus," *Dictionary of Scientific Biography*, vol. 4, ed. C. C. Gillispie (New York: Charles Scribner's Sons, 1971).

11. Larry Wright, "The Astronomy of Eudoxus: Geometry or Physics?" *Studies in History and Philosophy of Science* 4, no. 2 (1973): 165–172.

12. Aristotle, *On the Heavens*, trans. J. L. Stocks, Book 1, Part 5, Paragraph 4, http://classics.mit.edu/Aristotle/heavens.html.

13. Aristotle, *Heavens*, Book 1, Part 3, Paragraph 3.

14. Aristotle, *Heavens*, Book 1, Part 2, Paragraph 2.

15. Liba Chaia Taub, *Ptolemy's Universe* (Chicago: Open Court, 1993), 107–108.

16. Edward Grant, *Planets, Stars, & Orbs: The Medieval Cosmos, 1200–1687* (Cambridge: Cambridge University Press, 1994), 276.

17. Miguel A. Granada, "Aristotle, Copernicus, Bruno: Centrality, the Principle of Movement and the Extension of the Universe," *Studies in History and Philosophy of Science* 35, no. 1 (2004), 91.

18. Amelia Carolina Sparavigna, "Robert Grosseteste's Thought on Light and Form of the World," *International Journal of Sciences* 3, no. 4 (2014): 54–62.

19. Robert Grosseteste, *On Light,* trans. Claire C. Riedl (Milwaukee: Marquette University Press, 1942).

20. Yael Kedar and Giora Hon, "'Natures' and 'Laws': The Making of the Concept of Law of Nature—Robert Grosseteste (c. 1168–1253) and Roger Bacon (1214/1220–1292)," *Studies in History and Philosophy of Science* 61 (2017), 23.

21. Grosseteste, *On Light,* 10, 11.

22. Just exactly what the "firmament" was and how it fit into the medieval universe of spherical shells was an issue of much debate during the Middle Ages. For a good discussion, see Grant, *Planets, Stars, & Orbs,* 95–105.

23. Grosseteste, *On Light,* 14.

24. Kedar and Hon, "'Natures' and 'Laws,'" 24.

25. Crombie, *Robert Grosseteste,* 133.

26. Richard Bower, T. C. B. McLeish, B. K. Tanner, H. E. Smithson, C. Panti, N. Lewis, and G. E. M. Gasper, "A Medieval Multiverse? Mathematical Modelling of the Thirteenth Century Universe of Robert Grosseteste," *Proceedings of the Royal Society A* 470 (2014), 2.

27. Ten was considered a special number because, as the Pythagoreans of antiquity had noted, it is equal to one plus two plus three plus four. Grosseteste's 10 spheres apparently coincided with the firmament, the fixed stars, the five planets (Mercury, Venus, Mars, Jupiter, Saturn), the sun and moon, and the Earth itself (the sphere in the center).

28. Bower et al., "A Medieval Multiverse?" 8.

29. Bower et al., "A Medieval Multiverse?" 11.

30. McEvoy, *Robert Grosseteste,* 109.

Chapter 3. ARISTOTLE VERSUS THE ATOMISTS

Christiaan Huygens, *The Celestial Worlds Discover'd (Cosmotheoros): or, Conjectures Concerning the Inhabitants, Plants and Productions of the Worlds in the Planets* (London: Timothy Childe, 1698), 2–3.

1. St. Jerome (Hieronymus): Chronological Tables, 171st Olympiad, 171.3, http://www.attalus.org/translate/jerome2.html.

2. For brief accounts of what is known about Leucippus, see Sylvia Berryman, "Leucippus," *Stanford Encyclopedia of Philosophy,* Winter 2016 ed., Dec. 2, 2016, https://plato.stanford.edu/archives/win2016/entries/leucippus; and G. B. Kerferd, "Leucippus," *Dictionary of Scientific Biography,* vol. 8, ed. C. C. Gillispie (New York: Charles Scribner's Sons, 1973).

3. Steven Weinberg, *To Explain the World* (New York: Harper, 2015), 8.

4. John Burnet, *Early Greek Philosophy,* 3rd ed. (London: A & C Black Ltd., 1920), Chapter 9, Item 173.

5. G. S. Kirk, J. E. Raven, and M. Schofield, *The Presocratic Philosophers: A Critical History with a Selection of Texts,* 2nd ed. (Cambridge: Cambridge University Press, 1983), Fragment 549.

6. S. Sambursky, *The Physical World of the Greeks* (London: Routledge and Kegan Paul, 1963), 110–111.

7. This reasoning was better articulated by Lucretius. Accounts of Leucippus's views are too fragmentary to reveal the details of his motivations.

8. Kirk et al., *Presocratic Philosophers,* Fragment 563.

9. Burnet, *Greek Philosophy,* Chapter 9, Item 177.

10. The Pythagoreans, for instance, apparently believed that each star was a "world" with an Earth at its center.

11. Kirk et al., *Presocratic Philosophers,* comment after Fragment 565.

12. Kirk et al., *Presocratic Philosophers,* Fragment 565.

13. Sambursky, *Physical World,* 112.

14. A more recent brief account of the life of Epicurus is found in David J. Furley, "Epicurus," *Dictionary of Scientific Biography,* vol. 4, ed. C. C. Gillispie (New York: Charles Scribner's Sons, 1971).

15. As Meher Baba explained it, "The happiness of God-realization is self-sustained, eternally fresh and unfailing, boundless and indescribable. And it is for this happiness that the world has sprung into existence." Meher Baba, "Message of 1955," in *God Speaks: The Theme of Creation and Its Purpose* (New York: Dodd, Mead, 1973), 139.

16. Epicurus, *Letter to Herodotus,* in *The World of Physics,* ed. Jefferson Hane Weaver, 3 vols. (New York: Simon and Schuster, 1987), vol. 1, 310.

17. Epicurus, *Letter to Herodotus,* 1:320.

18. Sambursky, *Physical World,* 113.

19. See Tom Siegfried, "Finding a Quantum Way to Make Free Will Possible," *Science News* online, February 26, 2014, https://www

.sciencenews.org/blog/context/finding-quantum-way-make-free-will
-possible.

20. Epicurus, *Letter to Herodotus,* 1:311, 319.

21. David J. Furley, "Lucretius," *Dictionary of Scientific Biography,* vol. 8, ed. C. C. Gillispie (New York: Charles Scribner's Sons, 1973).

22. James H. Mantinband, "Introduction," in Lucretius, *On the Nature of the Universe,* trans. James. H. Mantinband (New York: Frederick Ungar, 1965), vi.

23. See, for instance, David Sedley, "Lucretius," *Stanford Encyclopedia of Philosophy,* Fall 2013 ed., August 10, 2013, https://plato.stanford.edu /archives/fall2013/entries/lucretius.

24. Lucretius, *On the Nature of Things,* trans. John Selby Watson, to which is appended the poetical version of John Mason Good (London: George Bell and Sons, 1880), 16–17.

25. Lucretius, *Nature of Things* (Watson translation), 49.

26. Lucretius, *Nature of Things* (Mason Good translation), 358.

27. Lucretius, *Nature of Things* (Watson translation), 93.

28. Sambursky, *Physical World,* 194.

29. Steven J. Dick, *Plurality of Worlds* (Cambridge: Cambridge University Press, 1982), 23.

Chapter 4. THE CONDEMNATION OF 1277

Pierre Duhem, *Le Système du Monde,* Book 6 (Paris: Hermann, 1954), 66. My translation.

1. The historian David Wootton, in *The Invention of Science* (New York: Harper, 2015), records the birth date of modern science as 1572, with the appearance of a nova, or "new star," observed by Tycho Brahe. The "invention" of science began in that year, Wootton contends, and ended with the publication of Newton's *Optics* in 1704.

2. Pierre Duhem, *Le Système du Monde,* Book 7 (Paris: Hermann, 1957), 4. Duhem credited the "birth certificate" phrasing to his friend and colleague Albert Dufourcq.

3. David C. Lindberg, *The Beginnings of Western Science,* 2nd ed. (Chicago: University of Chicago Press, 2007), 247.

4. As Steven Dick has pointed out, the importance of Tempier's edict in 1277 for the development of science is contentious, but "its effect on the idea of a plurality of worlds is unmistakable." See Steven J. Dick, *Plurality of Worlds* (Cambridge: Cambridge University Press, 1982), 28.

5. Plato, *Timaeus*, in *Theories of the Universe*, ed. Milton K. Munitz (Glencoe, IL: Free Press, 1957), 70.

6. Plato, *Timaeus*, 70.

7. See Jeremiah Hackett, "Bacon, Roger," *New Dictionary of Scientific Biography*, vol. 1, ed. Noretta Koertge (Detroit: Thomson Gale, 2008), 144; and Lindberg, *Beginnings*, 234. Hackett suggested that Bacon may only have met Grosseteste, and Lindberg wrote that while Bacon admired Grosseteste, he was "probably never his student." For further biographical information on Bacon, see A. C. Crombie and J. D. North, "Bacon, Roger," *Dictionary of Scientific Biography*, vol. 1, ed. C. C. Gillispie (New York: Charles Scribner's Sons, 1970).

8. Yael Kedar and Giora Hon, "'Natures' and 'Laws': The Making of the Concept of Law of Nature—Robert Grosseteste (c. 1168–1253) and Roger Bacon (1214/1220–1292)," *Studies in History and Philosophy of Science* 61 (2017), 21–31.

9. Kedar and Hon, "'Natures' and 'Laws,'" 22.

10. Francis Seymour Stevenson, *Robert Grosseteste, Bishop of Lincoln* (London: Macmillan and Co., 1899), 50–51.

11. Duhem, *Le Système du Monde*, Book 8, 33–34.

12. Duhem, *Le Système du Monde*, Book 8, 29. Scot also reiterated Aristotle's arguments that another cosmos would be impossible because it would require the element earth to move "naturally" in opposite directions—both toward the center of our world and toward the center of the other one.

13. Duhem, *Le Système du Monde*, Book 9, 366.

14. Duhem, *Le Système du Monde*, Book 9, 369.

15. For biographical background on Albert, see William Wallace, "Albertus Magnus," *Dictionary of Scientific Biography*, vol. 1, ed. C. C. Gillispie (New York: Charles Scribner's Sons, 1970); and Henryk Anzulewicz, "Albertus Magnus," *New Dictionary of Scientific Biography*, vol. 1, ed. Noretta Koertge (Detroit: Thomson Gale, 2008).

16. Lindberg, *Beginnings*, 239.

17. Lindberg, *Beginnings*, 238–239.

18. Dick, *Plurality of Worlds*, 37.

19. Duhem, *Le Système du Monde*, Book 9, 370.

20. Duhem, *Le Système du Monde*, Book 9, 371.

21. For more detail on Thomas, see Ralph McInerny and John O'Callaghan, "Saint Thomas Aquinas," *Stanford Encyclopedia of Philosophy*, Summer

2018 ed., May 23, 2014, https://plato.stanford.edu/archives/sum2018
/entries/aquinas.

22. Thomas Aquinas, *In libros Aristotelis De caelo et mundo exposition,* trans.
Fabian R. Larcher and Pierre H. Conway, Book 1, Lecture 16, 112,
http://dhspriory.org/thomas/DeCoelo.htm.

23. Aquinas, *In libros Aristotelis,* Book 1, Lecture 19.

24. Pierre Duhem, *Medieval Cosmology,* ed. and trans. Roger Ariew
(Chicago: University of Chicago Press, 1985), 449.

25. Michael J. Crowe, ed., *The Extraterrestrial Life Debate: Antiquity to 1915*
(Notre Dame, IN: University of Notre Dame Press, 2008), 19.

26. Edward Grant, *A History of Natural Philosophy* (Cambridge: Cambridge
University Press, 2007), 242.

27. Grant, *Natural Philosophy,* 245.

28. Edward Grant, ed., *A Source Book in Medieval Science* (Cambridge, MA:
Harvard University Press, 1974), 43.

29. Hans Thijssen, "Condemnation of 1277," *Stanford Encyclopedia of
Philosophy,* Winter 2016 ed., September 24, 2013, https://plato.stanford
.edu/archives/win2016/entries/condemnation.

30. Grant, *Source Book,* 47.

31. Grant, *Source Book,* 48–50.

32. Grant, *Source Book,* 46.

33. Lindberg, *Beginnings,* 247–248.

34. Grant, *Source Book,* 46–47.

35. Lindberg, *Beginnings,* 247–248.

Chapter 5. CONDEMNATION AFTERMATH

Edward Grant, *Planets, Stars, & Orbs: The Medieval Cosmos, 1200–1687*
(Cambridge: Cambridge University Press, 1994), 156.

1. Carl Sagan, *Cosmos* (New York: Random House, 1980), 556, 335.

2. Stefan Kirschner, "Nicole Oresme," *Stanford Encyclopedia of Philosophy,*
Fall 2017 ed., August 28, 2017, https://plato.stanford.edu/archives
/fall2017/entries/nicole-oresme.

3. Marshall Clagett, "Nicole Oresme and Medieval Scientific Thought,"
Proceedings of the American Philosophical Society 108, no. 4 (1964),
298.

4. Pierre Duhem, *Le Système du Monde,* Book 9 (Paris: Hermann, 1959),
380. My translation.

5. Edward Grant, *Planets, Stars, & Orbs: The Medieval Cosmos, 1200–1687*
(Cambridge: Cambridge University Press, 1994), 159.

6. Varon may have taught the influential philosopher John Duns Scotus (c. 1266–1308), aka Doctor Subtilis.

7. The translation for "Fundatus" is not obvious—it supposedly referred to the fact that he was a good teacher, and so might be translated as "thorough" or "established" or "capable."

8. Duhem, *Système du Monde,* Book 9, 381.

9. Frank Wilczek, "Multiversality," July 30, 2013, 2, arXiv:1307.7376.

10. Grant, *Planets, Stars, & Orbs,* 158.

11. Pasquale Porro, "Henry of Ghent," *Stanford Encyclopedia of Philosophy,* Fall 2014 ed., August 25, 2014, https://plato.stanford.edu/archives /fall2014/entries/henry-ghent.

12. Pierre Duhem, *Medieval Cosmology,* ed. and trans. Roger Ariew (Chicago: University of Chicago Press, 1985), 451.

13. Duhem, *Système du Monde,* Book 9, 375. Godfrey also rejected Aristotle's argument that the element earth would not know where to go if another world existed (the heaven surrounding the Earth would be a sufficient clue for matter to know which Earth it belonged to). And Godfrey disagreed that another world implied a void; another world would have "its own place," and there would be no void between the two worlds, for there would be no place there able to contain anything.

14. Duhem, *Medieval Cosmology,* 456–457.

15. Duhem, *Medieval Cosmology,* 470.

16. Edward Grant, *A History of Natural Philosophy* (Cambridge: Cambridge University Press, 2007), 183.

17. Grant, *Natural Philosophy,* 184.

18. Steven J. Dick, *Plurality of Worlds* (Cambridge: Cambridge University Press, 1982), 29.

19. Grant, *Planets, Stars, & Orbs,* 163n.

20. For an account of Ockham's life, see Philotheus Boehner, "Introduction," in William of Ockham, *Philosophical Writings,* trans. and ed. Philotheus Boehner, rev. with foreword by Stephen F. Brown (Indianapolis: Hackett, 1990). Another account of Ockham's life and times is William J. Courtenay, "The Academic and Intellectual Worlds of Ockham," in *The Cambridge Companion to Ockham,* ed. Paul Vincent Spade, 17–30 (Cambridge: Cambridge University Press, 1999). For further biographical information, see Ernest A. Moody, "Ockham, William of," *Dictionary of Scientific Biography,* vol. 10, ed. C. C. Gillispie (New York: Charles Scribner's Sons, 1974); and André Goddu, "Ockham, William of," *New Dictionary of Scientific Biography,* vol. 5, ed. Noretta Koertge (Detroit: Thomson Gale, 2008).

21. Boehner, "Introduction," xviii.

22. As paraphrased by Boehner, "Introduction," xviii.

23. Boehner, "Introduction," xxi.

24. William of Ockham, "What Is Motion?" trans. John E. Murdoch, in *A Source Book in Medieval Science,* ed. Edward Grant (Cambridge, MA: Harvard University Press, 1974), 230. The original Latin is *"Frustra fit per plura, quod potest fieri per pauciora."*

25. Isaac Newton, *Principia,* vol. 2: *The System of the World,* trans. Andrew Motte, rev. by Florian Cajori (Berkeley: University of California Press, 1934), 398.

26. George F. R. Ellis, "Does the Multiverse Really Exist?" *Scientific American* 305, no. 2 (August 2011), 43. In a later article, Ellis corrects his misquotation of Ockham's razor, rendering it as, "It is futile to do with more things that which can be done with fewer." See George Francis Rayner Ellis, "On the Philosophy of Cosmology," *Studies in History and Philosophy of Modern Physics* 46 (2014), 15.

27. Paul Davies, *Cosmic Jackpot* (Boston: Houghton Mifflin, 2007), 211.

28. Duhem, *Medieval Cosmology,* 463.

29. Duhem, *Medieval Cosmology,* 462.

30. Grant, *Planets, Stars, & Orbs,* 164.

31. James Franklin, *The Science of Conjecture: Evidence and Probability before Pascal* (Baltimore: Johns Hopkins University Press, 2001), 210.

32. Nicholas of Autrecourt (c. 1295–1369) was a French philosopher and theologian who held some controversial views, including belief in atoms. See Hans Thijssen, "Nicholas of Autrecourt," *Stanford Encyclopedia of Philosophy,* Spring 2016 ed., February 18, 2016, https://plato.stanford.edu/archives/spr2016/entries/autrecourt.

33. Nicole Oresme, *Le Livre du Ciel et du Monde,* ed. Albert D. Menut and Alexander J. Denomy, trans. Albert D. Menut (Madison: University of Wisconsin Press, 1968), 8.

34. Oresme's life and work are recounted in Marshall Clagett, "Oresme, Nicole," *Dictionary of Scientific Biography,* vol. 10, ed. C. C. Gillispie (New York: Charles Scribner's Sons, 1974); and Stefano Caroti, "Oresme, Nicole," *New Dictionary of Scientific Biography,* vol. 5, ed. Noretta Koertge (Detroit: Thomson Gale, 2008). See also Edward Grant, "Scientific Thought in Fourteenth-Century Paris: Jean Buridan and Nicole Oresme," *Annals of the New York Academy of Sciences* 314, no. 1 (1978): 105–126; and Kirschner, "Nicole Oresme."

35. Edward Grant, *Much Ado about Nothing: Theories of Space and Vacuum from the Middle Ages to the Scientific Revolution* (Cambridge: Cambridge University Press, 1981), 121.

36. Grant, *Much Ado,* 117–121.

37. Oresme, *Le Livre du Ciel,* 175–179.

38. Oresme, *Le Livre du Ciel,* 167.

39. Oresme, *Le Livre du Ciel,* 167. One later philosopher, John Major of Scotland (1467–1550), did appear to take the idea of concentric multiple worlds as a serious physical possibility. See Grant, *Planets, Stars, & Orbs,* 166–167.

40. Oresme, *Le Livre du Ciel,* 169, 171.

41. Oresme, *Le Livre du Ciel,* 177.

42. Grant, *Planets, Stars, & Orbs,* 166.

Chapter 6. CUSA AND COPERNICUS

E. H. Burritt, *The Geography of The Heavens* (Boston: Allen & Ticknor, 1833), 185.

1. Nicolaus Copernicus, *On the Revolutions of the Heavenly Spheres,* trans. Charles Glenn Wallis, Great Books of the Western World, vol. 16 (Chicago: Encyclopaedia Britannica, 1952), 516–517.

2. J. E. Hoffman, "Cusa, Nicholas," *Dictionary of Scientific Biography,* vol. 3, ed. C. C. Gillispie (New York: Charles Scribner's Sons, 1971), 513.

3. Cusa's career is described in Hoffman, "Cusa, Nicholas"; and Clyde Lee Miller, "Cusanus, Nicolaus [Nicolas of Cusa]," *Stanford Encyclopedia of Philosophy,* Summer 2017 ed., June 8, 2017, https://plato.stanford.edu /archives/sum2017/entries/cusanus.

4. Jasper Hopkins, *Nicholas of Cusa On Learned Ignorance: A Translation and an Appraisal of De Docta Ignorantia,* 2nd ed., (Minneapolis: Arthur J. Banning, 1985), Book 1, 6, 8.

5. Hoffman, "Cusa, Nicholas," 516.

6. Hopkins, *Nicholas of Cusa,* Preface, iv.

7. Hopkins, *Nicholas of Cusa,* Book 2, 70.

8. Hopkins, *Nicholas of Cusa,* Book 2, 80.

9. Nicholas of Cusa, *Of Learned Ignorance,* trans. Germain Heron (London: Routledge and Kegan Paul, 1954), quoted in *The Extraterrestrial Life Debate: Antiquity to 1915,* ed. Michael J. Crowe (Notre Dame, IN: University of Notre Dame Press, 2008), 29; and Hopkins, *Nicholas of Cusa,* Book 2, 94.

10. Cusa, *Learned Ignorance,* 31.

11. Hopkins, *Nicholas of Cusa,* Book 2, 110.

12. Cusa, *Learned Ignorance,* 32.

13. Steven J. Dick, *Plurality of Worlds* (Cambridge: Cambridge University Press, 1982), 41.

14. Dorothy Koenigsberger, "Decadence, Shift, Cultural Changes and the Universality of Leonardo da Vinci," in Phyllis Mack and Margaret C. Jacob, eds., *Politics and Culture in Early Modern Europe: Essays in Honor of H. G. Koenigsberger* (Cambridge: Cambridge University Press, 1987), 293. Cusa's influence on Leonardo may have been indirect; some experts suggest that Leonardo was more directly influenced by the Italian philosopher Marsilio Ficino.

15. Dick, *Plurality of Worlds,* 39–40.

16. For biographical information on Copernicus, see Edward Rosen, "Copernicus, Nicholas," *Dictionary of Scientific Biography,* vol. 3, ed. C. C. Gillispie (New York: Charles Scribner's Sons, 1971); and André Goddu, "Copernicus, Nicholas," *New Dictionary of Scientific Biography,* vol. 2, ed. Noretta Koertge (Detroit: Thomson Gale, 2008).

17. Owen Gingerich, "Did Copernicus Owe a Debt to Aristarchus?" *Journal for the History of Astronomy,* 16, no. 1 (1985), 40.

18. Nicolaus Copernicus, *De Revolutionibus (On the Revolutions),* trans. John F. Dobson and Selig Brodetsky, *Occasional Notes of the Royal Astronomical Society,* no. 10 (1947), in *Theories of the Universe,* ed. Milton Munitz (Glencoe, IL: Free Press, 1957), 156–157.

19. Copernicus, *On the Revolutions* (Wallis), 514.

20. Nicholas Copernicus, *On the Revolutions*, trans. Edward Rosen (Baltimore: Johns Hopkins University Press, 1992), 12.

21. Copernicus, *On the Revolutions* (Wallis), 519.

22. Copernicus *De Revolutionibus* (Dobson and Brodetsky), 164.

23. Copernicus, *On the Revolutions* (Wallis), 505.

Chapter 7. WANDERING IN IMMENSITY

Miguel Granada, "Kepler and Bruno on the Infinity of the Universe and of Solar Systems," *Journal for the History of Astronomy,* 39 (2008), 472.

1. Alexandre Koyré, *From the Closed World to the Infinite Universe* (Baltimore: Johns Hopkins Press, 1957), 54.

2. For biographical information on Thomas Digges, see Joy B. Easton, "Digges, Thomas," in *Dictionary of Scientific Biography,* vol. 4, ed. C. C.

Gillispie (New York: Charles Scribner's Sons, 1971); and André Goddu, "Digges, Thomas," *The Biographical Dictionary of Astronomers,* ed. Thomas Hockey (New York: Springer, 2007).

3. It's not clear whether Dee believed in the Copernican sun-centered system or merely accepted it as a method for mathematical calculations, as argued in J. Peter Zetterberg, "Hermetic Geocentricity: John Dee's Celestial Egg," *Isis* 70 (1979): 385–393.

4. It is not really relevant to the story, but for the record I'd like to mention that Field was my great-great-great-great-great-great-great-great-great grandfather.

5. Francis R. Johnson and Sanford V. Larkey, "Thomas Digges, the Copernican System, and the Idea of the Infinity of the Universe in 1576," *The Huntington Library Bulletin* No. 5 (April 1934), 109.

6. Owen Gingerich, *The Book Nobody Read* (New York: Walker, 2004), 119.

7. Johnson and Larkey, "Copernican System," 100.

8. Johnson and Larkey, "Copernican System," 88.

9. Johnson and Larkey, "Copernican System," 72.

10. Johnson and Larkey, "Copernican System," 105.

11. For accounts of Bruno's life and science, see Frances Yates, "Bruno, Giordano," *Dictionary of Scientific Biography,* vol. 2, ed. C. C. Gillispie (New York: Charles Scribner's Sons, 1970); and Enrico Giannetto, "Bruno, Giordano," *New Dictionary of Scientific Biography,* vol. 1, ed. Noretta Koertge (Detroit: Thomson Gale, 2008). See also Stillman Drake, "Copernicanism in Bruno, Kepler, and Galileo," *Vistas in Astronomy* 17 (1975), 180; and Miguel Granada, "Kepler and Bruno on the Infinity of the Universe and of Solar Systems," *Journal for the History of Astronomy* 39 (2008): 469–494.

12. I. Frith, *Life of Giordano Bruno the Nolan* (Boston: Ticknor & Co., 1887), 41.

13. Dorothea Waley Singer, *Giordano Bruno: His Life and Thought with Annotated Translation of His Work On the Infinite Universe and Worlds* (New York: Henry Schuman, 1950), 246.

14. Edward Grant, *Much Ado about Nothing: Theories of Space and Vacuum from the Middle Ages to the Scientific Revolution* (Cambridge: Cambridge University Press, 1981), 185.

15. Grant, *Much Ado,* 185.

16. Koyré, *Closed World,* 36.

17. Francis Johnson, "Astronomical Thought in Renaissance England," in *Theories of the Universe,* ed. Milton Munitz (Glencoe, IL: Free Press, 1957), 189.

18. Drake, "Copernicanism," 180.
19. Koyré, *Closed World,* 39.
20. Granada, "Kepler and Bruno," 470.
21. Singer, *Giordano Bruno,* 231.
22. Granada, "Kepler and Bruno," 470, 477. A key point is that Bruno did not merely propose an infinite space containing countless stars surrounding the central sun and Earth, but that the universe has no center; all the stars, including the sun, are on equal footing, each surrounded by its own planets.
23. Singer, *Giordano Bruno,* 302.
24. Frith, *Life of Giordano Bruno,* 44, 46.
25. Koyré, *Closed World,* 44.
26. Singer, *Giordano Bruno,* 250–251.
27. Singer, *Giordano Bruno,* 251, 304.
28. J. V. Field, *Kepler's Geometrical Cosmology* (Chicago: University of Chicago Press, 1988), 18.
29. For a thorough account of Kepler's controversial and colorful life, see Max Caspar, *Kepler,* trans. C. Doris Hellman (New York: Dover, 1993). For an assessment of his life and work see Walter Gerlach, "Johannes Kepler—Life, Man and Work," *Vistas in Astronomy* 18 (1975): 73–95 (plus many other entries in that volume). For biographical accounts of his science, see Owen Gingerich, "Kepler, Johannes," *Dictionary of Scientific Biography,* vol. 7, ed. C. C. Gillispie (New York: Charles Scribner's Sons, 1973); James R. Voelkel, "Kepler, Johannes," *New Dictionary of Scientific Biography,* vol. 4, ed. Noretta Koertge (Detroit: Thomson Gale, 2008); and Adam Jared Apt, "Kepler, Johannes," *The Biographical Dictionary of Astronomers,* ed. Thomas Hockey (New York: Springer, 2007).
30. Johannes Kepler, *De Stella Nova* (Prague: P. Sessii, 1606), 106. My translation.
31. Koyré, *Closed World,* 62.
32. Koyré, *Closed World,* 78.
33. Johannes Kepler, *Epitome Astronomiae Copernicanae* (Johannes Plancus, 1618), 34. My translation.
34. Dennis Richard Danielson, *The Book of the Cosmos* (Cambridge, MA: Perseus, 2000), 171. Or, in another translation, "the primary bosom of the universe." See Granada, "Kepler and Bruno," 482.
35. Kepler, *De Stella Nova,* 109.
36. Field, *Kepler's Geometrical Cosmology,* 18.
37. Kepler, *Epitome,* 40.
38. Koyré, *Closed World,* 70.

39. Alberto Martinez, "Giordano Bruno and the Heresy of Many Worlds," *Annals of Science* 73 (2016): 345–374.

40. Galileo Galilei, *Dialogue Concerning the Two Chief World Systems,* trans. Stillman Drake, 2nd ed. (Berkeley: University of California Press, 1967), 319.

41. A. C. Crombie, *Science, Optics and Music in Medieval and Early Modern Thought* (London: Habledon Press, 1990), 343.

42. Steven J. Dick, *Plurality of Worlds* (Cambridge: Cambridge University Press, 1982), 44.

43. The story of this find is told in Stephen Greenblatt, *The Swerve: How the World Became Modern* (New York: Norton, 2011).

44. Dick, *Plurality of Worlds,* 46.

45. Michel de Montaigne, *The Complete Essays of Montaigne,* trans. Donald M. Frame (Stanford, CA: Stanford University Press, 1965), 390.

46. Montaigne, *Complete Essays,* 390.

47. For information on Gassendi's life, see Carla Rita Palmerino, "Gassendi, Pierre," *New Dictionary of Scientific Biography,* vol. 3, ed. Noretta Koertge (Detroit: Thomson Gale, 2008); and Bernard Rochot, "Gassendi (Gassend), Pierre," *Dictionary of Scientific Biography,* vol. 5, ed. C. C. Gillispie (New York: Charles Scribner's Sons, 1972).

48. Grant, *Much Ado,* 207–210.

49. A point made by Dick, *Plurality of Worlds,* 54.

50. Dick, *Plurality of Worlds,* 56.

51. Walter Charleton, *Physiologia Epicuro-Gassendo-Charltoniana, or, A fabrick of science natural, upon the hypothesis of atoms founded by Epicurus repaired [by] Petrus Gassendus* (London: Thomas Heath, 1654), 12–13.

52. Henry More, *Democritus Platonissans* (1646), The Augustan Reprint Society, Pub. no. 131, William Andrews Clark Memorial Library, University of California, Los Angeles, 1968), "Introduction" by P. G. Stanwood, "To the Reader," and verses 20 and 21, Project Gutenberg eBook, https://ia800503.us.archive.org/11/items/democritus platon30327gut/30327-0.txt.

Chapter 8. PLANETS AND PEOPLE

Bernard le Bovier de Fontenelle, *A Plurality of Worlds,* trans. John Glanvill (London: R. Bentley and S. Magnes, 1688), 126.

1. Arthur Conan Doyle, *A Study in Scarlet* (1887), in *The Complete Sherlock Holmes* (Garden City, NY: Doubleday, 1930), 21.

2. In later editions, Fontenelle added a sixth evening.

3. One of many biographies of Descartes is A. C. Grayling, *Descartes: The Life and Times of a Genius* (New York: Walker, 2005). For more information on his life and philosophy, see Gary Hatfield, "René Descartes," *Stanford Encyclopedia of Philosophy*, Summer 2018 ed., January 16, 2014, https://plato.stanford.edu/archives/sum2018/entries /descartes. For further information on his life and science see Daniel Garber, "Descartes, René du Perron," *New Dictionary of Scientific Biography*, vol. 2, ed. Noretta Koertge (Detroit: Thomson Gale, 2008); and A. C. Crombie, "Descartes, René du Perron," *Dictionary of Scientific Biography*, vol. 4, ed. C. C. Gillispie (New York: Charles Scribner's Sons, 1971) and the appended account on Descartes's mathematics and physics by Michael S. Mahoney.

4. Hatfield, "René Descartes."

5. René Descartes, *The Principles of Philosophy*, trans. John Veitch, 1901, (Kindle edition, Amazon Digital Services, 2011), 72.

6. Alexandre Koyré, *From the Closed World to the Infinite Universe* (Baltimore: Johns Hopkins Press, 1957), 104, 109. Koyré suggested that the common opinion of the time held that the indefinite-infinite issue was a pseudo-distinction, made for the purpose of placating the theologians.

7. Koyré, *Closed World*, 104.

8. Descartes, *Principles of Philosophy*, 72.

9. René Descartes, *Oeuvres de Descartes*, vol. 4 (Paris: Chez F. G. Levrault for Victor Cousin, 1824), 270. My translation.

10. Descartes, *Principles of Philosophy*, 75.

11. Steven Weinberg, *To Explain the World* (New York: Harper, 2015), 204.

12. Lucía Ayala, "Worlds and Systems in Early Modern Europe," January 30, 2013, 4, arXiv:1301.7317v1.

13. Steven J. Dick, *Plurality of Worlds* (Cambridge: Cambridge University Press, 1982), 141.

14. For a discussion of Fontenelle's life, see Nina Rattner Gelbart, "Introduction," in Bernard le Bovier de Fontenelle, *Conversations on the Plurality of Worlds*, trans. H. A. Hargreaves (Berkeley: University of California Press, 1990); and Suzanne Delorme, "Fontenelle, Bernard le Bouyer (or Bovier) de," *Dictionary of Scientific Biography*, vol. 5, ed. C. C. Gillispie (New York: Charles Scribner's Sons, 1972).

15. Bernard le Bovier de Fontenelle, *Entretiens sur la Pluralité des Mondes, augmentés des dialogues des morts* (Lyon: De l'Imprimerie de Leroy, 1800), v–vi. My translation.

16. Dick, *Plurality of Worlds,* 126.

17. Fontenelle, *Entretiens,* iv, vii.

18. Fontenelle, *Entretiens,* 6.

19. Fontenelle, *Entretiens,* 17.

20. Fontenelle, *Entretiens,* 50–51.

21. Fontenelle, *A Plurality of Worlds,* trans. John Glanvill (London: R. Bentley and S. Magnes, 1688), 86.

22. Fontenelle, *Conversations on the Plurality of Worlds,* trans. H. A. Hargreaves (Berkeley: University of California Press, 1990), 49.

23. Fontenelle, *Entretiens,* 93–94.

24. Fontenelle, *Entretiens,* 94.

25. Fontenelle, *Entretiens,* 94–96.

26. Fontenelle, *Plurality* (Glanvill), 124–125.

27. Fontenelle, *Plurality* (Glanvill), 125–126.

28. Fontenelle, *Plurality* (Glanvill), 151–152.

29. Fontenelle, *Plurality* (Glanvill), 8.

30. Fontenelle, *Plurality* (Glanvill), 64–65.

31. Fontenelle, *Entretiens,* 7–8.

32. Fontenelle, *Entretiens,* 55.

33. For biographical information on Huygens, see H. J. M. Bos, "Huygens, Christiaan," *Dictionary of Scientific Biography,* vol. 6, ed. C. C. Gillispie (New York: Charles Scribner's Sons, 1972); and Henk Kubbinga, "Huygens, Christiaan," *The Biographical Dictionary of Astronomers,* ed. Thomas Hockey (New York: Springer, 2007).

34. Christiaan Huygens, *The Celestial Worlds Discover'd (Cosmotheoros): or, Conjectures Concerning the Inhabitants, Plants and Productions of the Worlds in the Planets* (London: Timothy Childe, 1698), 1–8.

35. Huygens, *Celestial Worlds,* 18–19.

36. Huygens, *Celestial Worlds,* 80.

37. Huygens, *Celestial Worlds,* 145, 149, 150–151.

38. Huygens, *Celestial Worlds,* 156.

39. I. Bernard Cohen, ed., *Isaac Newton's Papers and Letters on Natural Philosophy* (Cambridge, MA: Harvard University Press, 1958), 281–283.

40. Cohen, *Isaac Newton's Papers,* 284, 287.

41. Isaac Newton, *Opticks* (New York: Dover, 1952), 404.

Chapter 9. ISLAND UNIVERSES

Edgar A. Poe, *Eureka: A Prose Poem* (New York: G. P. Putnam, 1848), 97–98.

1. Bernard le Bovier de Fontenelle, *Conversations on the Plurality of Worlds,* trans. H. A. Hargreaves (Berkeley: University of California Press, 1990), 66.

2. Martin Schönfeld and Michael Thompson, "Kant's Philosophical Development," *Stanford Encyclopedia of Philosophy,* Winter 2014 ed., November 25, 2014, https://plato.stanford.edu/archives/win2014 /entries/kant-development.

3. For information on Wright's life, see Michael Hoskin, "The Cosmology of Thomas Wright of Durham," *Journal for the History of Astronomy* 1, no. 1 (1970): 44–52; Vera Gushee, "Thomas Wright of Durham, Astronomer," *Isis* 33, no. 2 (1941): 197–218; and Hoskin, "Wright, Thomas," *Dictionary of Scientific Biography,* vol. 14, ed. C. C. Gillispie (New York: Charles Scribner's Sons, 1976).

4. Hoskin, "Cosmology of Thomas Wright," 44.

5. Michael J. Crowe, *The Extraterrestrial Life Debate: 1750–1900* (1986; Mineola, NY: Dover, 1999), 42.

6. Thomas Wright, *The Universe and the Stars,* first American edition, from the London edition of 1750, with notes by C. S. Rafinesque (Philadelphia: for C. Wetherill, 1837), 9.

7. Wright, *The Universe and the Stars,* 104–105.

8. Wright, *The Universe and the Stars,* 104, 117, 131.

9. At least that was one of Wright's suggestions. He mentioned other possibilities, and later in life appeared to change his mind about some of his original ideas about the cosmos.

10. Hoskin, "Cosmology of Thomas Wright"; Gushee, "Thomas Wright of Durham."

11. Lucía Ayala, "Worlds and Systems in Early Modern Europe," January 30, 2013, 8, arXiv:1301.7317v1.

12. Wright, *The Universe and the Stars,* 143–144.

13. For an account of Kant's life and science see James W. Ellington, "Kant, Immanuel," *Dictionary of Scientific Biography,* vol. 7, ed. C. C. Gillispie (New York: Charles Scribner's Sons, 1973). For information on his life and philosophy, see Michael Rohlf, "Immanuel Kant," *Stanford Encyclopedia of Philosophy,* Summer 2018 ed., January 25, 2016, https://plato.stanford.edu/archives/sum2018/entries/kant; and Schönfeld and Thompson, "Kant's Philosophical Development."

14. G. J. Whitrow, "Kant and the Extragalactic Nebulae," *Quarterly Journal of the Royal Astronomical Society* 8 (1967), 53.

15. Whitrow, "Kant and the Extragalactic Nebulae," 53.

16. Immanuel Kant, *Allgemeine Naturgeschichte und Theorie des Himmels [Universal Natural History and Theory of the Heavens]*, in *Kant's Cosmogony*, ed. and trans. W. Hastie (Glasgow: James Maclehose and Sons, 1900), 56.

17. Kant, *Allgemeine Naturgeschichte*, 63.

18. Crowe, *Extraterrestrial Life Debate, 1750–1900*, 48.

19. Michael J. Crowe, ed., *The Extraterrestrial Life Debate: Antiquity to 1915* (Notre Dame, IN: University of Notre Dame Press, 2008), 163.

20. Herschel's life and science are discussed in M. A. Hoskin, "Herschel, William," *Dictionary of Scientific Biography*, vol. 6, ed. C. C. Gillispie (New York: Charles Scribner's Sons, 1972); and in Michael J. Crowe and Keith R. Lafortune, "Herschel, (Friedrich) William [Wilhelm]," *The Biographical Dictionary of Astronomers*, ed. Thomas Hockey (New York: Springer, 2007).

21. Hector Macpherson, *Herschel* (New York: Macmillan Company, 1919), 49.

22. Macpherson, *Herschel*, 54.

23. Ayala, "Worlds and Systems," 9.

24. Edwin Hubble, *The Realm of the Nebulae* (New Haven: Yale University Press, 1936), 25.

25. Robert Smith, *The Expanding Universe: Astronomy's 'Great Debate' 1900–1931* (Cambridge: Cambridge University Press, 1982), 48.

26. Biographical information about Mitchel can be found in Trudy E. Bell, "Mitchel, Ormsby MacKnight," *The Biographical Dictionary of Astronomers*, ed. Thomas Hockey (New York: Springer, 2007); P. C. Headley, *Old Stars: The Life and Military Career of Major-General Ormsby M. Mitchel* (Boston: Lee and Shepard, 1883); "General Mitchell," *Scientific American* 7 (November 22, 1862), 329–330; and F. A. Mitchel, *Ormsby MacKnight Mitchel: Astronomer and General* (Boston: Houghton, Mifflin and Co., 1887).

27. O. M. Mitchel, *The Planetary and Stellar Worlds* (New York: Baker and Scribner, 1848), 324. Actually, he referred to "our own 'island universe'" and "the thousands of these astral systems that exist in space."

28. *Huron Reflector* (Norwalk, OH), April 13, 1847.

29. *The Evening Post* (New York), December 18, 1847.

30. *Sidereal Messenger* 1, No. 5 (October 1846).

31. Alexander von Humboldt, *Cosmos*, vol. 3, trans. Edward Sabine (London: Longman, Brown, Green, and Longmans, 1852), 227.

32. David N. Stamos, *Edgar Allan Poe,* Eureka, *and Scientific Imagination* (Albany: State University of New York, 2017).

33. I discussed Poe's anticipation of the Big Bang theory in more detail in my book *Strange Matters* (Washington, DC: Joseph Henry Press, 2002). While there are similarities to modern cosmology, Poe's ideas do not coincide precisely with today's theories and lacked technical mathematical detail.

34. Edgar A. Poe, *Eureka: A Prose Poem* (New York: G. P. Putnam, 1848), 96, 101.

35. Poe, *Eureka,* 102.

36. Kant, *Allgemeine Naturgeschichte,* 64–65.

37. Kant, *Allgemeine Naturgeschichte,* 65.

Chapter 10. E PLURIBUS UNIVERSE

A. C. D. Crommelin, "Are the spiral nebulae external galaxies?" *Journal of the Royal Astronomical Society of Canada* 12 (February 1918), 46. Reprinted from *Scientia,* May 1917.

1. Robert Smith, presentation at the April meeting of the American Physical Society, Washington, DC, January 28, 2017.

2. *Huron Reflector,* April 13, 1847.

3. *Sidereal Messenger,* October 1847, 22.

4. For more biographical information on Huggins, see Herbert Dingle, "Huggins, William," *Dictionary of Scientific Biography,* vol. 6, ed. C. C. Gillispie (New York: Charles Scribner's Sons, 1972); Mary T. Brück, "Huggins, William," *The Biographical Dictionary of Astronomers,* ed. Thomas Hockey (New York: Springer, 2007); and Barbara J. Becker, "Eclecticism, Opportunism, and the Evolution of a New Research Agenda: William and Margaret Huggins and the Origins of Astrophysics" (PhD diss., Johns Hopkins University, 1993), https://faculty.humanities.uci.edu/bjbecker/huggins.

5. Sir William Huggins and Lady Huggins, eds., *The Scientific Papers of Sir William Huggins* (London: William Wesley and Son, 1909), 498.

6. Huggins, *Scientific Papers,* 118–119.

7. Virginia Trimble, "Multiverses of the Past," *Astronomische Nachrichten* 330, no. 7 (2009), 767.

8. J. Lévy, "The Exploration of the Stellar Universe," in *Science in the Nineteenth Century,* ed. René Taton, trans. A. J. Pomerans (New York: Basic Books, 1965), 129.

9. Agnes Clerke, *The System of the Stars* (London: Longmans, Green and Co., 1890), 368. Clerke acknowledged that the "practical certainty" that the Milky Way constituted the entire universe was justified "so far as our capacities of knowledge extend." Perhaps more existed than we could know, but she declared that suspicion irrelevant to astronomy. "With the infinite possibilities beyond," she wrote, "science has no concern." Some of today's multiverse deniers echo that attitude.

10. Agnes Clerke, *A Popular History of Astronomy during the Nineteenth Century,* 3rd ed. (London: Adam & Charles Black, 1893), 505.

11. J. Gore, "The Sidereal Heavens," in *The Concise Knowledge Astronomy,* ed. A. Clerke, J. Gore, and A. Fowler (London: Hutchinson and Co., 1898), 545–546, quoted in Robert Smith, *The Expanding Universe: Astronomy's 'Great Debate' 1900–1931* (Cambridge: Cambridge University Press, 1982), 16.

12. J. Ellard Gore, *Studies in Astronomy* (London: Chatto and Windus, 1904), 137.

13. Stefano Bettini, "A Cosmic Archipelago: Multiverse Scenarios in the History of Modern Cosmology," October 12, 2005, arXiv:physics /0510111.

14. Arthur Berry, *A Short History of Astronomy from Earliest Times through the Nineteenth Century* (1898; New York: Dover, 1961), 405–406.

15. The "April" meeting retains its name because it refers to the physics subfields that are represented, in contrast to other subfields that are represented at the "March" meeting, which is almost always held in March (although it was in February once).

16. For more on Keeler, see Glenn A. Walsh, "Keeler, James Edward," *The Biographical Dictionary of Astronomers,* ed. Thomas Hockey (New York: Springer, 2007); and Sally H. Dieke, "Keeler, James Edward," *Dictionary of Scientific Biography,* vol. 7, ed. C. C. Gillispie (New York: Charles Scribner's Sons, 1973).

17. Smith, *The Expanding Universe,* 26.

18. A. C. D. Crommelin, "Are the Spiral Nebulae External Galaxies?" *Journal of the Royal Astronomical Society of Canada* 12 (February 1918), 45. Reprinted from *Scientia,* 11, no. 21 (1917).

19. Harlow Shapley, "A Faint Nova in the Nebula of Andromeda," *Publications of the Astronomical Society of the Pacific* 29, no. 171 (1917): 213–217.

20. For more on Curtis, see Jordan D. Marché II and Rudi Paul Lindner, "Curtis, Heber Doust," *The Biographical Dictionary of Astronomers,* ed. Thomas Hockey (New York: Springer, 2007); Michael A. Hoskin,

"Curtis, Heber Doust," *Dictionary of Scientific Biography,* vol. 3, ed. C. C. Gillispie (New York: Charles Scribner's Sons, 1971); and Robert G. Aitken, "Heber Doust Curtis," *Biographical Memoirs of the National Academy of Sciences* 22 (1943): 275–294.

21. For more on Shapley, see Horace A. Smith and Virginia Trimble, "Shapley, Harlow," *The Biographical Dictionary of Astronomers,* ed. Thomas Hockey (New York: Springer, 2007); Owen Gingerich, "Shapley, Harlow," *Dictionary of Scientific Biography,* vol. 12, ed. C. C. Gillispie (New York: Charles Scribner's Sons, 1975); and Bart J. Bok, "Harlow Shapley," *Biographical Memoirs of the National Academy of Sciences* 49 (1978): 241–291.

22. For a thorough account of the Shapley-Curtis debate, see Michael A. Hoskin, "The 'Great Debate': What Really Happened," *Journal for the History of Astronomy* 7 (1976): 169–182. See also Smith, *The Expanding Universe,* 77–88.

23. For more on Hubble, see Helge Kragh, "Hubble, Edwin Powell," *The Biographical Dictionary of Astronomers,* ed. Thomas Hockey (New York: Springer, 2007); G. J. Whitrow, "Hubble, Edwin Powell," *Dictionary of Scientific Biography,* vol. 6, ed. C. C. Gillispie (New York: Charles Scribner's Sons, 1972); and N. U. Mayall, "Edwin Powell Hubble," *Biographical Memoirs of the National Academy of Sciences* 41 (1970): 175–214.

24. For more on Leavitt, see Harry G. Lang, "Leavitt, Henrietta Swan," *The Biographical Dictionary of Astronomers,* ed. Thomas Hockey (New York: Springer, 2007); and Owen Gingerich, "Leavitt, Henrietta Swan," *Dictionary of Scientific Biography,* vol. 6, ed. C. C. Gillispie (New York: Charles Scribner's Sons, 1973).

25. Not surprisingly, it's a little more complicated than that, because it turns out that not all Cepheids are alike—a source of some early confusion. And it was a little tricky to determine the distance to any Cepheid to begin with for the calibration (none were close enough to use parallax), although that was accomplished by the Danish astronomer Ejnar Hertzsprung soon after Leavitt discovered the period-luminosity relationship.

26. *Science News-Letter* 5, December 6, 1924. A similar report appeared in the *New York Times,* November 23, 1924.

27. *Science News-Letter*, December 6, 1924.

28. For a more thorough and entertaining discussion of who discovered Hubble's distance-velocity law, see Virginia Trimble, "Anybody but Hubble!" July 8, 2013, arXiv:1307.2289.

29. Edwin Hubble and Milton L. Humason, "The Velocity-Distance Relationship among Extra-Galactic Nebulae," *Astrophysical Journal* 74 (1931): 43–80.

30. Edwin Hubble, "A Relation between Distance and Radial Velocity among Extra-Galactic Nebulae," *Proceedings of the National Academy of Sciences* 15 (1929), 173.

31. Edwin Hubble, *The Realm of the Nebulae* (New Haven: Yale University Press, 1936), 122.

32. For a thorough discussion of all the factors that led to the triumph of the Big Bang theory, see Helge Kragh, *Cosmology and Controversy* (Princeton: Princeton University Press, 1996).

Chapter 11. MANY QUANTUM WORLDS

Frank Wilczek, "Multiversality," July 28, 2013, 7, arXiv:1307.7376.

1. *Star Trek,* "The Tholian Web," directed by Herb Wallerstein, written by Gene Roddenberry et al., Paramount Television, November 15, 1968.

2. *Star Trek: The Next Generation,* "Parallels," directed by Robert Wiemer, written by Brannon Braga, Paramount Television, December 12, 1993.

3. Max Jammer, *The Philosophy of Quantum Mechanics* (New York: John Wiley and Sons, 1974), 517.

4. The photoelectric effect is the phenomenon by which light induces the emission of electrons from certain materials.

5. Some physicists and writers quibble with this description, arguing that it's meaningless to say an electron (or anything else) exists in multiple places at once. But it's difficult to convey the curious nature of a superposition of states in any other way that makes much sense, either. I think it's better to say the electron exists in multiple locations than to say it exists nowhere. The key point is that it is absolutely wrong to say that it occupies a definite position that is merely unknown before an observation or measurement.

6. Aage Petersen, "The Philosophy of Niels Bohr," *Bulletin of the Atomic Scientists* 19, no. 7 (1963), 12. This statement has been widely quoted and sometimes ridiculed by Bohr's critics, but it does not reflect the full depth of Bohr's views, which he articulated at greater length in various talks and papers. A more complete statement was that "in our description of nature the purpose is not to disclose the real essence of the phenomena but only to track down, so far as it is possible, relations between the manifold aspects of our experience." See Niels Bohr, *The*

Philosophical Writings of Niels Bohr, vol. 1: *Atomic Theory and the Description of Nature* (1934; Woodbridge, CT: Ox Bow Press, 1987), 18. The importance and implications of this point have been discussed in N. David Mermin, "Making Better Sense of Quantum Mechanics," September 5, 2018, 18–20, arXiv:1809.01639.

7. John A. Wheeler, *Geons, Black Holes, and Quantum Foam* (New York: W. W. Norton, 1998), 268. For a thorough biography of Everett, see Peter Byrne, *The Many Worlds of Hugh Everett III* (Oxford: Oxford University Press, 2010).

8. Wheeler, *Geons,* 268.

9. Hugh Everett III, "The Theory of the Universal Wave Function," in *The Many-Worlds Interpretation of Quantum Mechanics,* ed. Bryce DeWitt and Neill Graham (Princeton: Princeton University Press, 1973), 6.

10. Everett, "Universal Wave Function," 107.

11. Everett's expression of the compatibility of his idea with Bohr's views was no doubt included at the urging of Wheeler, who heavily edited Everett's thesis. See Byrne, *The Many Worlds of Hugh Everett III.*

12. Paul C. Aichelburg and Roman U. Sexl, eds., *Albert Einstein: His Influence on Physics, Philosophy and Politics* (Braunschweig / Wiesbaden: Friedr. Vieweg & Sohn, 1979), 202.

13. Everett, "Universal Wave Function," 116.

14. Everett, "Universal Wave Function," 116–117.

15. Hugh Everett III, *The Everett Interpretation of Quantum Mechanics,* ed. Jeffrey A. Barrett and Peter Byrne (Princeton: Princeton University Press, 2012), 274.

16. Everett, "Universal Wave Function," 117.

17. Hugh Everett III, "'Relative State' Formulation of Quantum Mechanics," *Reviews of Modern Physics* 29, no. 3 (1957): 458–460.

18. Wheeler, *Geons,* 268.

19. John A. Wheeler, "Assessment of Everett's 'Relative State' Formulation of Quantum Theory," *Reviews of Modern Physics* 29, no. 3 (1957), 465, 464.

20. Everett, *Everett Interpretation,* 274, 276.

21. Everett, *Everett Interpretation,* 11.

22. Bryce DeWitt, presentation at Science & Ultimate Reality conference, Plainsboro, NJ, March 16, 2002.

23. Bryce DeWitt, "Quantum Mechanics and Reality," *Physics Today* 23, no. 9 (1970): 30–35. Reprinted in DeWitt and Graham, eds., *The Many-Worlds Interpretation,* 155.

24. DeWitt, "Quantum Mechanics," 161, 163.

25. Wheeler, *Geons,* 270.

26. Tom Siegfried, "Feline Physics," *Dallas Morning News,* October 19, 1987.

27. DeWitt, presentation at Science & Ultimate Reality conference.

28. Siegfried, "Feline Physics."

29. Siegfried, "Feline Physics."

30. Charles Bennett, email message to author, June 15, 1994.

31. Wojciech Zurek, "Decoherence and the Transition from Quantum to Classical," *Physics Today* 44 (1991): 36–44. Zurek posted an updated version of this paper in 2002 at https://arxiv.org/abs/quant-ph/0306072.

32. Tom Siegfried, "Reality Check," *Dallas Morning News,* April 1, 2002.

33. Murray Gell-Mann and James Hartle, "Adaptive Coarse Graining, Environment, Strong Decoherence, and Quasiclassical Realms," *Physical Review A* 89, no. 5 (2014), 052125, 2.

34. Murray Gell-Mann and James Hartle, "Quantum Mechanics in the Light of Quantum Cosmology," in *Complexity, Entropy and the Physics of Information,* ed. W. H. Zurek (Redwood City, CA: Addison-Wesley, 1990), 430.

35. Gell-Mann and Hartle, "Adaptive Coarse Graining," 1.

36. Gell-Mann and Hartle, "Adaptive Coarse Graining," 1.

37. Murray Gell-Mann, presentation at the annual meeting of the American Association for the Advancement of Science, Chicago, February 11, 1992.

38. Murray Gell-Mann, interview with the author, August 21, 2009, Princeton, NJ.

39. This "ontic" versus "epistemic" issue is very complicated and contentious. See Tom Siegfried, "Physicists Debate Whether Quantum Math Is as Real as Atoms," *Science News* online, January 15, 2015, https://www.sciencenews.org/blog/context/physicists-debate-whether-quantum-math-real-atoms.

40. Tom Siegfried, "Quantum Mysteries Dissolve If Possibilities Are Realities," *Science News* online, October 1, 2017, https://www.sciencenews.org/blog/context/quantum-mysteries-dissolve-if-possibilities-are-realities.

41. Tom Siegfried, "'QBists' Tackle Quantum Problems by Adding a Subjective Aspect to Science," *Science News* online, January 15, 2014, https://www.sciencenews.org/blog/context/qbists-tackle-quantum-problems-adding-subjective-aspect-science.

Chapter 12. ANTHROPIC COSMOLOGY

Joe Polchinski, interview with author in Santa Barbara, CA, May 9, 2006.

1. *Galen on the Usefulness of the Parts of the Body,* trans. Margaret Talladge May (Ithaca, NY: Cornell University Press, 1968), 189–191.

2. For a deep investigation of the history of anthropic reasoning as it relates to cosmology, see Stefano Bettini, "Anthropic Reasoning in Cosmology: A Historical Perspective," October 19, 2004, arXiv:physics/0410144. A well-known book treating anthropic cosmology is John Barrow and Frank Tipler, *The Anthropic Cosmological Principle* (New York: Oxford University Press, 1986).

3. Steven Weinberg, public lecture, Cleveland, OH, October 9, 2003.

4. Joe Lykken, conversation with author, October 11, 2003, Cleveland, OH.

5. Andrei Linde, interview with author, January 7, 1991, The Woodlands, TX.

6. Alice Calaprice, *The Ultimate Quotable Einstein* (Princeton: Princeton University Press, 2011), 344. Linde used a slightly different translation.

7. Andrei Linde, "Inflation and Quantum Cosmology," in *300 Years of Gravitation,* ed. Stephen Hawking and Werner Israel (Cambridge: Cambridge University Press, 1987), 607.

8. Andrei Linde, "Eternally Existing Self-Reproducing Chaotic Inflationary Universe," *Physics Letters B* 175, no. 4 (1986), 399.

9. Linde, "Eternally Existing," 399.

10. Martin Rees, *Before the Beginning* (Reading, MA: Perseus Books, 1997), 3.

11. Steven Weinberg, "Anthropic Bound on the Cosmological Constant," *Physical Review Letters* 59, no. 22 (1987): 2607–2610.

12. Steven Weinberg, "The Cosmological Constant Problem," *Reviews of Modern Physics* 61, no. 2 (1989), 20.

13. Steven Weinberg, panel discussion, Kavli-Cerca Cosmology Conference, Cleveland, OH, October 10, 2003.

14. Weinberg, public lecture.

15. The celebration was several months late, Wheeler having been born in 1911. Technically it was held in Plainsboro, New Jersey, at a conference center just outside Princeton.

16. Being "independent" means that these other universes in a Level II multiverse would never be reachable by any spaceship (or signal) traveling at the speed of light. It might be possible, on the other hand, to visit other "universes" in our Level I multiverse if cosmic

acceleration someday slows and reverses. That's maybe not a good bet, but it's theoretically conceivable. For details see Max Tegmark, "The Multiverse Hierarchy," May 8, 2009, arXiv:0905.1283 and his book *The Mathematical Universe* (New York: Random House, 2014).

17. Jaume Garriga, Alexander Vilenkin, and Jun Zhang, "Black Holes and the Multiverse," February 25, 2016, 12, 38, arXiv:1512.01819v4. These authors point out that in their scenario, black holes with a wide range of masses would be produced near the end of inflation. If future observations find that the number of black holes of various masses agrees with the prediction based on their calculations, "it could be regarded as evidence for inflation and for the existence of a multiverse."

18. Max Tegmark, presentation at the Science & Ultimate Reality conference, Plainsboro, NJ, March 16, 2002.

19. Andrei Linde, presentation at the Science & Ultimate Reality conference, Plainsboro, NJ, March 17, 2002.

20. David Gross, panel discussion, at Kavli-Cerca Cosmology Conference, Cleveland, OH, October 11, 2003.

21. Steven Weinberg, interview with author, October 19, 1989, Austin, TX.

22. Polchinski, interview.

23. Shamit Kachru, Renata Kallosh, Andrei Linde, and Sandip P. Trivedi, "de Sitter Vacua in String Theory," *Physical Review D* 68, no. 4 (2003), 046005.

24. S. W. Hawking and Thomas Hertog, "Populating the Landscape: A Top-Down Approach," *Physical Review D* 73 (2006), 123527.

25. The 14th International Conference on Supersymmetry and the Unification of Fundamental Interactions, Newport Beach, CA, June 12–17, 2006.

26. Burton Richter, "Randall and Susskind," letter to editor, *New York Times*, January 29, 2006. Panel discussion, Conference on Supersymmetry, Newport Beach, CA, June 14, 2006.

27. Leonard Susskind, interview with author, June 14, 2006, Newport Beach, CA.

28. Polchinski, interview.

29. Susskind, interview.

30. David Gross, telephone interview with author, June 5, 2006.

31. Sean Carroll, telephone interview with author, June 8, 2006.

Chapter 13. BRANE WORLDS

H. G. Wells, *Men Like Gods* (New York: The Macmillan Co., 1923), 49.

1. Wells, *Men Like Gods*, 55.

2. Francisco Caruso and Roberto Xavier make a case that Kant did not really explain why space has three dimensions but rather why objects can extend in only three dimensions. See Francisco Caruso and Roberto Moreira Xavier, "On Kant's First Insight into the Problem of Space Dimensionality and Its Physical Foundations," April 25, 2015, arXiv:0907.3531v2.

3. Caruso and Xavier, "Kant's First Insight," 8.

4. Edwin A. Abbott, *Flatland: A Romance of Many Dimensions* (London: Seeley and Co., 1884; repr. New York: Dover, 1992).

5. Simon Newcomb, "Modern Mathematical Thought," address delivered before the New York Mathematical Society at the annual meeting, December 28, 1893, *Bulletin of the New York Mathematical Society* 3 (January 1894), 105.

6. Newcomb, "Modern Mathematical Thought," 105–106.

7. Charles Hinton, *Scientific Romances. No. 1. What is the Fourth Dimension?* (London: W. Swan Sonnenschein & Co., 1884), 23, 19.

8. Hinton, *Scientific Romances,* 19, 24.

9. Hinton, *Scientific Romances,* 26.

10. Hinton, *Scientific Romances,* 27, 31.

11. Another scientist who contemplated an extra dimension was Karl Pearson (1857–1936), who suggested that atoms were created by "squirts of ether" into space from outside our three dimensions. Karl Pearson, *The Grammar of Science* (1892; London: J. M. Dent & Sons, 1937), 226–229.

12. Tom Siegfried, "Superstrings Snap Back," *Dallas Morning News,* March 19, 1990.

13. Nima Arkani-Hamed, Savas Dimopoulos, and Gia Dvali, "The Hierarchy Problem and New Dimensions at a Millimeter," *Physics Letters B* 429, no. 3 (1998): 263–272.

14. Savas Dimopoulos, presentation at the Seventh International Conference on Supersymmetries in Physics, Fermilab, June 17, 1999.

15. Lisa Randall, presentation at the Seventh International Conference on Supersymmetries in Physics, Fermilab, June 17, 1999.

16. Randall, presentation at the Seventh International Conference on Supersymmetries in Physics.

17. Lisa Randall, *Warped Passages* (New York: HarperCollins, 2005), 7, 61.
18. Randall, *Warped Passages,* 61.
19. Extradimensional brane worlds also suggest the possibility of another kind of multiverse: universes recurring in time, like the ancient Stoics believed (as mentioned in chapter 5). In fact, physicists including Paul Steinhardt, Roger Penrose, and others have advocated a view of this sort to explain our known universe without recourse to inflationary cosmology. Such "cyclic multiverses" are an interesting possibility, but most physicists do not now consider them to be as worthy of consideration as inflation. "The Cyclic Multiverse is widely known within the physics community but is viewed, almost as widely, with much skepticism," Brian Greene wrote in *The Hidden Reality* (New York: Alfred A. Knopf, 2011), 124.

Chapter 14. DEFINING THE MULTIVERSE

Fred Hoyle, "The Future of Physics and Astronomy," *American Scientist* 64 (March–April 1976), 202.

1. William James, *Is Life Worth Living?* (Philadelphia: S. Burns Weston, 1896), 21–22, 26.
2. Webster's Third New International Dictionary, Unabridged (Springfield, MA: Merriam-Webster, 1993), 1486.
3. *The Age* (Melbourne, Australia), November 26, 1904.
4. *Eau Claire Leader* (Eau Claire, WI), February 16, 1915.
5. *The Brooklyn Daily Eagle* (Brooklyn, NY), September 17, 1931.
6. Even before inflation was proposed, some physicists had discussed the possibility of multiple universes, although without (as far as I have been able to tell) using the term *multiverse*. For a thorough account of the early discussions of the idea, see Stefano Bettini, "A Cosmic Archipelago: Multiverse Scenarios in the History of Modern Cosmology," October 12, 2005, arXiv:physics/0510111.
7. Jack Williamson, "Kinsman to Lizards," *Analog Science Fiction / Science Fact* (July 1978), 92. Science fiction writer Michael Moorcock seems to have used the term *multiverse* earlier, but his multiverse resembled Everett's many quantum worlds more than the multiverse of inflationary cosmology.
8. Max Tegmark, *The Mathematical Universe* (New York: Random House, 2014). In a historical survey of various multiverse ideas, Virginia Trimble offers a somewhat different categorization of multiverse levels.

See Virginia Trimble, "Multiverses of the Past," *Astronomische Nachrichten* 330, no. 7 (2009): 761–769.

9. Brian Greene, *The Hidden Reality* (New York: Alfred A. Knopf, 2011), 309.

10. A common objection to the universe as simulation is that there would be no way to know if that were the case. But some investigators have suggested possible tests. See Tom Siegfried, "Maybe There's a Way to Find Out If Reality Is a Computer Simulation," *Science News Prime,* December 10, 2012, https://www.sciencenews.org/article/tom-siegfried -randomness-14.

11. Frank Wilczek, "Multiversality," July 28, 2013, 3, arXiv:1307.7376.

12. Edgar A. Poe, *Eureka: A Prose Poem* (New York: G. P. Putnam, 1848), 103.

13. Sean Carroll, "Beyond Falsifiability: Normal Science in a Multiverse," January 15, 2018, 2, arXiv:1801.05016.

14. Steven Weinberg, interview with author, February 15, 2018, Austin, TX.

15. Weinberg, interview.

16. John Donoghue, interview with author, March 23, 2018, Amherst, MA.

17. V. Agrawal, S. M. Barr, John F. Donoghue, and D. Seckel, "Anthropic Considerations in Multiple-Domain Theories and the Scale of Electroweak Symmetry Breaking," *Physical Review Letters* 80, no. 9 (1998): 1822–1825.

18. Weinberg, interview.

19. James Hartle, "Quantum Multiverses," January 25, 2018, 24, arXiv:1801.08631v1.

20. Hartle, "Quantum Multiverses," 29.

21. Hartle, "Quantum Multiverses," 4.

22. Wilczek, "Multiversality," 8–9.

23. Hartle, "Quantum Multiverses," 34.

24. Hartle, "Quantum Multiverses," 36.

25. Weinberg, interview.

26. Isaac Newton, *Principia,* vol. 2: *The System of the World,* trans. Andrew Motte, rev. by Florian Cajori (Berkeley: University of California Press, 1934), 398.

27. While it seems very unlikely that other universes could be directly observed, it is possible in principle that one bubble universe could impinge on our bubble and generate a detectable pattern of temperature anomalies in the cosmic microwave background radiation. There have been hints of such anomalies, but nothing definitive. Nevertheless statements that the multiverse is not scientific because other universes

are not observable in principle are simply not necessarily true. See Diana Steele, "When Worlds Collide," *Science News* 173, no. 18 (2008): 22–25.

28. George Francis Rayner Ellis, "On the Philosophy of Cosmology," *Studies in History and Philosophy of Modern Physics* 46 (2014), 14–15.

29. Ellis, "Philosophy of Cosmology," 15.

30. Ellis, "Philosophy of Cosmology," 17.

31. Weinberg, interview.

32. Donoghue, interview.

33. Wilczek, "Multiversality," 2.

34. Sean Carroll, telephone interview with author, June 8, 2006.

35. Carroll, "Beyond Falsifiability," 4, 12.

36. Donoghue, interview.

37. Frank Wilczek, interview with author, June 13, 2006, Newport Beach, CA.

38. Weinberg, interview.

39. Donoghue, interview.

40. Martin Rees, public lecture at the April meeting of the American Physical Society, Washington, DC, January 28, 2017.

EPILOGUE

Lucretius, *Nature of Things,* trans. John Mason Good (London: George Bell and Sons, 1880), 357.

1. Arvind Borde, Alan H. Guth, and Alexander Vilenkin, "Inflationary Spacetimes Are Incomplete in Past Directions," *Physical Review Letters* 90, no. 15 (2003), 151301.

2. "In any kind of inflating cosmology," Leonard Susskind writes, "the odds strongly favor the beginning to be so far in the past that it is effectively at minus infinity." Susskind, "Is Eternal Inflation Past-Eternal? And What If It Is?" May 3, 2012, arXiv:1205.0689. See also Tom Siegfried, "Bubble Universes Give New Perspective to Time's Origin and Its Arrow," *Science News Prime,* June 25, 2012, https://www.sciencenews .org/article/bubble-universes-give-new-perspective-times-origin-and -its-arrow.

3. A. C. Crombie, *The History of Science from Augustine to Galileo,* vol. 2 (1959; New York: Dover, 1995), 68.

4. Mario Livio and Martin Rees, "Fine-Tuning, Complexity, and Life in the Multiverse," January 22, 2018, 13, arXiv:1801.06944.

5. It's curious that today many scientists seem to endorse Karl Popper's philosophical requirement that proper theories must be falsifiable, despite the fact that his view is no longer favored by philosophers. Sean Carroll quotes the philosopher Alex Broadbent on this point: "Popper remains extremely popular among natural scientists, despite almost universal agreement among philosophers that—notwithstanding his ingenuity and philosophical prowess—his central claims are false." See Carroll, "Beyond Falsifiability: Normal Science in a Multiverse," January 15, 2018, 4, arXiv:1801.05016.

6. Paul Steinhardt, presentation at the Texas Symposium on Relativistic Astrophysics, Dallas, TX, December 9, 2013.

7. Helge Kragh, *Cosmology and Controversy* (Princeton: Princeton University Press, 1996), 175.

8. Alan Guth, "Quantum Fluctuations in Cosmology and How They Lead to a Multiverse," December 27, 2013, 1, arXiv:1312.7340.

9. John Donoghue, "The Multiverse and Particle Physics," *Annual Review of Nuclear and Particle Science* 66 (2016): 1–21; Frank Wilczek, "Multiversality," July 28, 2013, arXiv:1307.7376.

10. Brian Greene, *The Hidden Reality* (New York: Alfred A. Knopf, 2011), 318.

11. Joe Polchinski, interview with author, May 9, 2006, Santa Barbara, CA.

12. Steven Weinberg, interview with author, February 15, 2018, Austin, TX.

13. John P. A. Ioannidis, "Why Most Published Research Findings Are False," *PLoS Medicine* 2, no. 8 (2005), e124.

14. Steven Weinberg, public lecture in Cleveland, OH, October 9, 2003.

15. Leonard Susskind, interview with author, June 14, 2006, Newport Beach, CA.

16. René Descartes, *The Principles of Philosophy,* trans. John Veitch (1901), Kindle, 75.

17. Christiaan Huygens, *The Celestial Worlds Discover'd (Cosmotheoros): or, Conjectures Concerning the Inhabitants, Plants and Productions of the Worlds in the Planets* (London: Timothy Childe, 1698), 9.

18. Huygens, *Celestial Worlds,* 69.

Illustration Credits

Index

Page numbers in *italics* represent images.